# From Down Beat to Vinyl:

## Bill Putnam's Legacy To The Recording Industry

By Bob Bushnell and Jerry Ferree

# From Down Beat to Vinyl:

## Bill Putnam's Legacy To The Recording Industry

By Bob Bushnell and Jerry Ferree

Cover Art: Craig Dulmes, In Decent Exposure Studios

All Photographs by the Authors

www.bookstandpublishing.com

Published by
Bookstand Publishing
Morgan Hill, CA 95037
3273_4



All comments and opinions in this book are those of the authors only. Although we have the greatest faith in Milton T. Putnam's son, Bill Putnam and Universal Audio, his company; he had no financial nor editorial hand in the writing of this book.

ISBN 978-1-58909-830-5

Printed in the United States of America

# Dedication

We dedicate this book to our former employer and mentor, and to Bud Morris and Tony Parri, two fascinating people who figured prominently in the story of Bill's many successes and triumphs. Sadly, all have now passed away. We also wish to thank all the engineers, mixers and individuals that we worked with, and to all the composers, arrangers, conductors, musicians, singers and recording artists who provided the music, a major reason for the studios and the industry.

Bob: To my wife, Olga, the best friend I ever had. To my daughters, and my former wife, who tolerated (?) all the late nights.

Jerry: To my wife, Louanne, who put up with the long hours while we were empire building and my children who rarely saw me except on weekends. And, of course, to Bill without whom this adventure and this book would not exist.

# About the Authors

Although we're both retired, we stay busy from year to year. Visiting family and grandchildren, keeping up appearances, clearing up the east forest, remodeling our homes, seeing good friends; all those things that truly make retirement an enjoyment. (By the way, we're serious about that!)

If questions arise, or consultations desired, our email address is:

mtpbf@bresnan.net or abcjerry@presys.com

We will be glad to answer questions about the book, and the recording business.

Jerry lives in Oregon, Bob lives in Montana. (But they don't argue about which is the best place to live. They both are right.)

# Preface

**Bob:** Much has been written about Bill. Mix Magazine has published two biographies about Bill, albeit with some errors. The latest was written in 2003.

Bill's real name was Milton Tasker Putnam. He mentioned that he had gotten the nickname of Bill when he was much younger, and it stayed with him. Once he showed me a sales letter addressed to him as "Mr. William Putnam." He was amused, but scornful.

Bill was married four times and had four children; a son and a daughter by his first wife, and two sons by his third wife.

Bill was an engineer and mixer first, then a businessman. He was very aware of the value of the money, but he didn't put money before his art. To him, I believe it was art, not just what he did for a living, and he loved it.

He was impulsive, no question about that. Some projects he did through the years could have had more advance planning, but then, that wasn't Bill.

Bill got his start in the business when he and his partners started a network delay facility to record East Coast network programs on transcription discs and play them back on the line later, thereby keeping programs at the same relative time across the country. They operated out of a converted storefront in Palatine, Illinois a northwestern suburb of Chicago.

Bill's first studio in Chicago (as far as we know) was on the 8th floor of the Chicago Opera House at 20 N. Wacker Drive. The building is now called the Civic Opera House.

It's well known that Bill got his start as a producer by producing and releasing a record of "Peg O' My Heart," by Jerry Murad and the Harmonicats. He recorded that when the studio was at 20 N. Wacker Drive. They used the men's room as an echo chamber. The record was released on Bill's own label, Universal Records. (Not the original Universal Records with the thick 78-rpm records that had a ¾ inch hole in the center).

**Jerry:** Bill was also known for his acoustical engineering prowess and his ingenuity for solving all kinds of audio related problems.

He developed the "cascode" preamplifier circuitry for audio, and used it in his successful tube amplifiers, and a new twist on using feedback to control distortion in a limiting amplifier. He was a master at jury-rigging a quick fix when problems arose during a recording session. He practically wrote the book on multiple microphone placement to minimize cross pickup from other parts of the room.

**Jerry:** He helped to pioneer multiple recording techniques such as those used by Les Paul and Mary Ford, his good friends. He wrote songs with his 2nd wife Belinda, recorded them and released them on his own label.

Bill lived in an analog world. He never saw digital recording equipment, and never had any training in digital techniques. He would have been thoroughly delighted. Knowing Bill as we did, we know that he would have begun a crash course to catch up and master the new techniques as soon as possible. That was Bill.

**Bob:** He came up with the idea of tape reverb, which was possible as the Ampex machines had a separate playback head. He would simply introduce some of the recorded signal back into the recording. It was great for mono, but not so good for stereo.

Jerry remarked that Bill's world was analog. However, when transistors and solid-state electronics appeared, rather than downgrading the technology, Bill eagerly embraced the technology. He set up evening classes at United for those engineers who wanted to learn more about transistors. We bought our own textbooks.

After this foreword we've added a glossary for those unfamiliar with recording terms. At the back of the book, you'll find a list of the many people and engineers we worked with, our version of the history of recording, and Web links to sites about recording. The history is from our understanding, and might be argued by some individuals. So be it.

## A word (or two) to those working in the Recording Business:

**Bob**: Jerry and I started in the business when you learned by doing. No schools, just paying attention. Making mistakes along the way, putting your ingenuity, intelligence and common sense to work. Ingenuity, intelligence and common sense are still basic requirements, and mistakes will still be made.

Even though we're somewhat older, and don't have the ears we did back then, we still have an appreciation for the quality of most of today's music. We're also quietly appalled by what passes for music in some circles.

We're also impressed by the number of new individuals pursuing this fascinating business. The large studios, such as Radio Recorders, United and Western, Coastal Recording in New York, and Universal Recording in Chicago have literally shrunk or disappeared as the style of recording has drastically changed. Film scoring is alive and well, both in the U.S. and Europe. As Jerry later discusses, the nature of film scoring requires stages or studios dedicated to that purpose.

Even though this book was written on two computers, about 800 miles apart, and the manuscript was sent back and forth via email, and later on a SmartMedia disk; back then we worked in an analog world, where the word "digital" wasn't even a zero on the horizon. The only 'computers' were monstrous machines taking up hundreds of square feet and with relatively little ability judging by today's standards.

It is fascinating that in this world of project studios, digital workstations, and very small digital field recorders, analog isn't dead. Analog mike preamps, and analog limiters are still a part of the recording world. The Fairchild 670 limiter, much larger than a Universal Audio 1176, still is a sought-after beast. We used both of those devices.

In the digital world, in an effort to improve quality, sampling rates, first at 44.1 kHz, have increased steadily. Digital is always sampling, however, analog is continuous.

An interesting item: a tremendous blending of two wildly different formats, to put it mildly. A father and son team in Germany, together with a conductor and an orchestra have created an astounding CD of the famed singer, Enrico Caruso. They used the original 78-rpm records, digitally removed the noise, and the original orchestra, then had the conductor and orchestra *underdub* Caruso's arias. They used tempo tracks which they built to match the original tempi. The result is magnificent, and just a trifle spooky.

Among other things, this CD is an excellent reference for the quality of recording circa 1925, and the use of computers and the appropriate software today. It also demonstrates the mutual cooperation and respect that musicians and engineers have to this day.

# Glossary

Normally, a glossary is at the end, however some readers may not understand all the terms used. We thought it would be better at the beginning.

**Acetate**: Another term for the lacquer coated blank used to make master disc recordings and dubs.

**Agency**: Short for Advertising Agency. Common usage in the industry.

**Analog**: The technique by which we humans talk and listen. All LP records are an analog recording. All natural or human sounds in the real world are analog.

**Board**: Another term for a recording console. A mixer is usually considered a portable device.

**Cardioid**: Used with microphones, one of several patterns describing the pickup pattern of the microphone. So-called, because the pattern is roughly heart shaped.

**Console**: An elaborate device in front of which the engineer sits, controlling and equalizing the sounds from the various microphones in the studio. Also used for mix-down work.

**Condenser** (or capacitor) microphone: A type of microphone whereby sound vibrations will move one of two small diaphragms that are very close together (but never touching). The two diaphragms form the plates of a capacitor and are part of an electronic circuit.

**Cover record**: Term for a song recorded by one artist and copied by another artist. The second artist was equally or better known than the first artist. The intent was to capitalize on the popularity of the second artist. Not in current use, because of the play lists generated by broadcasting syndicates.

**Digital**: A formatted code of 0s and 1s strung together in a continuous stream. How information is stored on a computer, how information is sent on the Internet. How information is stored on a CD or DVD.

**Dub**: A record made on a record blank intended for immediate playback. The record blank was essentially the same as those used for masters, but of lower quality.

**Duplicator system**: A system whereby a master tape is copied to many slave tapes at once. Originally designed for reel-to-reel tapes, and later for Compact Cassette tapes.

**Echo chamber**: A room, approximately 1000 cubic feet in size, wherein all surfaces are highly acoustically reflective, and none are parallel. A loudspeaker and microphone (usually non-directional) are placed in the chamber so that the loudspeaker doesn't face the microphone. Electromechanical reverb devices began to replace echo chambers in the early 1960's. Solid state digital reverb devices began to appear in the late 70's and rapidly gained popularity due to their small size and easy portability.

**Editing Session**: A process, by which the engineer, usually under the direction of the producer, physically edits or cuts the magnetic tape. This is done to remove or change a problem that occurred during the recording, or to combine two takes to make a tune without apparent errors. With the advent of multi-track tape and later digital and hard disk recording, this process is unnecessary. Nowadays, the engineer will perform the editing using a computer instead of scissors.

**Equalizer**: An electronic device to elaborately affect the tonal characteristic of the signal being fed through it.

**EQ'd**: Alluding to the fact that an audio signal has been electronically altered in some way to enhance its aural properties.

**Fader**: An industry term for the volume control, one for each microphone. Originally designed as rotary units, later on they were linear in action. So-called, as they were used to fade or lower the microphone volume.

**HP-LP Filter**: A passive device for sharply limiting the upper and lower frequency response of an audio signal.

**Lacquer**: Another term for the record blank. This was a circular disc of aluminum, very flat, coated with a proprietary lacquer based compound, usually dark purple in color. The stylus in the cutter-head would cut a groove in the relatively soft coating, usually averaging about 2 thousandths of an inch in width.

**Lathe**, or recording lathe: A complex precision mechanical device, central to which is a turntable rotating at the correct speed, 33 1/3, 45 or 78 rpm. On the turntable rests the record blank. The lathe controls the precise horizontal movement of the cutter-head across the record blank, creating a long spiral groove. On stereo records the groove varies in width and depth.

**LED**: Short for Light Emitting Diode. Used primarily as indicators in electronic equipment. They didn't appear until the late 60's. Now widely used in TV and computer screens, replacing cathode ray tubes.

**Light pattern measurements**: A process whereby a collimated or pinpoint light beam is reflected across a master test lacquer for accurately checking the frequency response of the analog cutter head system. Also known as Buchmann-Meyer patterns.

**Limiter**: An electronic device to automatically control the level of the audio being fed through the device. Also sometimes referred to as a "Compressor."

**Mixer**: A small, sometimes portable audio console. Also, one who mixes or operates a console.

**Mixing or Mix**: The process wherein the engineer controls the sound levels and quality from the various microphones during a recording session.

**Mastering**: A process whereby the audio material is transferred to a recording lathe that cuts the lacquer master from which the LP or 45 rpm dies are prepared. These dies are used to manufacture vinyl records for sale.

**Mix-down**: A process whereby the recorded material is remixed or combined into two stereo tracks for phonograph records or CD's and four or more tracks for films.

**Overdub (and underdub)**: A process whereby a live vocal or instrument is recorded on a second tape machine, together with the original music recording. Underdub is the opposite whereby live music is recorded on a second tape machine, together with the original vocal. Both techniques became obsolete once multi-track machines came into common use, with rare exceptions.

**Pink noise**: Filter shaped random noise used for acoustical measurements. This definition is necessarily brief.

**Pot or slider**: A synonymous term for a fader.

**Repeat coil**: An audio transformer (usually 600 ohms impedance for both primary and secondary windings) used to electrically isolate two circuits. Very seldom used nowadays.

**Reverberation**: A term used in acoustics to describe part of the characteristics of a room or studio. In recording, frequently shortened to "reverb" or "echo", it applies to part of an audio signal being fed to an echo chamber then combined with the balance of the material. The term 'reverb' was often used as a verb, as in "Add some reverb to the guitar."

**Ribbon microphone**: Referring to the microphone design, in which a ribbon or ribbons are stretched within a magnetic field. The sound vibrations cause the ribbon to move, thus generating an electrical signal.

**Stylus**: Referring to both a playback stylus in a phonograph cartridge and a sapphire stylus used in cutting dubs or lacquer masters.

**Tape reverb**: A process, used during recording or mix-down, whereby a part of the playback signal from the tape being recorded is fed back to combine with the other signals being fed through the console. This creates a repeating effect.

**VU meter**: An analog measuring device showing the amount of audio signal on a continuous basis. Its reading is usually weighted to allow a more average reading. Now primarily replaced by LED strips in many mixing consoles. This (LED) type is usually a peak reading device, and is particularly useful for digital recording.

**Williamson type amplifier**: A high quality vacuum tube power amplifier, known for its low harmonic distortion and very popular in the 50's and 60's. This type of amplifier is returning to the high-end audiophile market for music lovers still skeptical of digital recording and solid state equipment.

# The Adventure Begins

## Ontario St.
## 1953

Bob: In high school, I went to work at a local radio repair shop. At the high school, we had a small recording studio, with an 8 input board. Two Presto 6N disk recorders with Presto cutter heads were the recording system. My math teacher, Bob Anspaugh, who was in charge of the studio, was the individual who started me off in the intricacies of recording. I later worked for another radio repair shop. I recall buying some of the first LPs, that were invented and produced by Columbia Records.

Two high school buddies and I experimented with electronics after school. After high school, I went to work for an electronics distributor, Allied Radio, in Chicago, which enabled me to buy a Magnecord PT-6 monaural tape recorder at a discounted price.

With all that behind me, it was hardly surprising that I ended up in the recording business. As you will read, Jerry had a similar background.

My first meeting with Bill Putnam was at a meeting of the Chicago Audio and Acoustical Group sometime in late 1953. Bill's studio, Universal Recording Corporation, 111 E. Ontario in Chicago was the meeting place. I attended as a neophyte in audio. Approaching Bill after the meeting, I asked about working at the studio. He suggested I drop by some evening, as he didn't work during the day. It seemed strange at the time, but after the first evening it was clear. Back then, R&B and pop sessions didn't happen during the day.

I went to see Bill one evening after work. I don't remember the conversation at all, probably through nervousness. He suggested that I work evenings for awhile and see "how things would work out." I worked during the day at an electronics distributor, then drove just north of the Loop to the studio. I'd work from about 7 PM to around midnight, drive home (I was living with my parents), grab some sleep and back to work the next morning.

After all, I was only 22, and sleep didn't mean very much to me then. I was hired as a maintenance engineer though what I knew about recording equipment was practically nil. Much to my delight, Bill suggested I work full-time for him starting January 1954. I gave my notice to the electronics distributor, which meant I was now working just one job.

## 1954

When I started, the tape machines were practically brand-new flat top Ampex 300's, machines using 10 ½ inch reels. For some reason, which I never asked about, we always used 7-inch reels.

**Jerry:** The tape was Scotch 111A, which was the only formulation available then. It came on 7" plastic reels, aluminum hubs (no flanges and difficult to handle) and aluminum reels. I asked why we used the 7" plastic reels, and was told that they were cheaper and that they usually were plenty long enough to record "spots" or "jingles", the staple of our business.

**Bob:** Ampex first came out with the 200, then the 400, then came the 300, then the 350 and 351. These were all mono, of course, in 1954. One of my first tasks was to replace the VU meters with better quality units. I remember rolling a converted 300 to Studio A; then at a break, Bill would help me get it up to the control room. It was up a short flight of stairs, as the control room was about three feet above the studio floor. Those Ampex 300 flattops weren't light!

When Bill was mixing a session, he'd drop the input gain on the Ampex by 4 or 6 dB. It was easy to do, as Ampex was using Daven rotary stepped attenuators, calibrated in 1 dB steps. He'd let the meter on the board peg itself most of the time, but he'd watch the meter on the Ampex for real levels. Because of the limited range of the VU meter, he was effectively extending the meter range. That's one idea I've never forgotten.

Bill's partners, Bernie Clapper and Bob Weber, did agency dates and mastering through the day. Bill did all kinds of music for record companies (usually at night and on weekends). Universal had no regular maintenance staff (except me), so if anything quit working, Bernie or Bob just fixed the problem and went on from there. When I got there at night, I'd spend some time in going through the studios and workrooms gathering up tools. Apparently, that habit has stuck with me, as I still make a point of keeping tools together.

**Bob:** I recall one real goof I made. Bill, Stan White (more about him later) and I were modifying a Western Electric radio console for use in Studio B as a recording board. Built in true WE style, the console was desktop, manufactured with 14-gauge steel, rather heavy. The filament cables were shielded, and were not small. My task was to drill various holes to support cable clamps and wires. I wasn't watching what I was doing, just leaning on the drill motor. Suddenly, I got through the side, but right through one of the filament wires as well. It took most of the evening to repair my goof. Bill reminded me about that for years afterwards.

The building layout for Ontario St. was somewhat unusual. Walking in the front door, past Ampex's branch office, then past the receptionist; if you turned left, down the hall, you came to Bill and Belinda's office. Belinda was Bill's second wife. They had a small black dog, named Jet, who followed Belinda around the studio unless he was sent back to the office. If you turned right, there was a small office, which Tony Parri later occupied.

Straight down the hall, you passed the dubbing room, about which you'll hear more later. After that was the maintenance shop, big enough for the workbench and maybe three people. Then still on your right, were the stairs up to the tape library. Straight on was the door to Studio A. If you turned left before you got to the studio, you went to Studio B and the mastering room.

The restroom locations were somewhat unusual. To get to them, you went through Studio A to the far left corner; there was a door to a small hallway, and the two restrooms. How did you use the restrooms during a session? You waited for a break!

The glamorous design of the studios we see today hadn't yet arrived on the scene. Stereo hadn't appeared either, so with mono recording, loudspeaker placement was less critical. In Studio A, the monitor was to the left, just over the tape machines. I recall Bill usually mixing with his head turned to the left. That way, he had the same sound in both ears.

Bill had designed Studio A using poly-cylindrical baffles with absorptive baffles hung from the ceiling. Not a glamorous design, but very functional. The floor was asphalt tile, some beige color. All of the mike receptacles with Hubbell Twist-Lok connectors were under the control room window, with short runs to the board, since the studio was much wider than it was deep. The lights were fluorescent fixtures. As I remember, Bill had used remote ballasts even then to cut down on the ambient noise. Remember, we're talking about 1954.

I spent a fair amount of time in the control room, watching and listening to Bill as he mixed. After listening and watching, I realized just how good were his understanding of music and the producer's desires.

Studio B was a much smaller room. Bernie Clapper usually used the control room for editing ABC radio shows. He used a Magnecord PT6 with rack mount ears for 10 1/2 inch reels. The room itself wasn't very large.

I remember mixing a R&B session for a small label in Chicago. I don't remember whether I volunteered, or Bill asked me to do the date. It was a small session, with three or four musicians. The producer, who was also the label owner, was drinking gin and smoking a large cigar. The combination finally got to me, and I made a hasty trip to the restroom. I came back and apologized, but I couldn't finish the session.

I came back to work the next night, but I don't recall Bill ever saying anything. I'm sure the studio didn't charge the client, however. I don't know if that was the reason, but I didn't want to do any more mixing until a few years later.

Next door was the mastering room, which had two Scully lathes with Cook cutter-heads. The room was functional, no glamour. One lathe was 33 or 45, and the other was 33 or 78. Presto amps that Bill had very extensively modified to accommodate the Cook heads were mounted above the lathes. The Cook cutter-heads used feedback to help flatten response and reduce distortion. They weren't motional feedback (Westrex 2A mono cutter-heads were the first motional feedback design). Later, Fairchild designed a stereo motional feedback cutter, which Bill used in Los Angeles.

The board in Studio A had ten inputs, with no eq for each input. In other words, his mixes were final. No mixdown, just basic eq and limiting when making dubs or mastering. When Bill asked for help on setup (we didn't have setup boys back then), I did my share of setups and breakdowns. Remembering inputs and mikes wasn't a problem. Vocal on 1, bass on 2, piano on 3, guitar on 4, drums on 5, woodwinds on 6, brass on 7. As you see, he grouped the rhythm section together, and the brass and woodwinds together. He was one of the first mixers to use tape reverb. The tape machine return was on 8 or 10, so it was easy for him to bring up the tape output as necessary. When three-track recording came along, creating stereo records, the technique didn't work as well.

These were analog monaural machines and consoles, so overdubbing was inevitably a loss of one generation. But with all the tapes of Bill's that I heard and worked with, the levels were good, never a problem. As for mixdowns: they didn't exist. "We'll fix it in the mix"; that phrase didn't come along until around 1965. If the mixer blew a take, and he was friends with the producer, no problem. Bill seldom blew a take. Bill wasn't simply a mixer; he understood the music as well. More than once, I remember Bill suggesting a particular rhythm to the client. He related to the clients and musicians. In later years, half in jest, half serious, producing charges were added to the rate card. But I don't think they ever charged a client for production time.

The Ampex 3200 duplicator system was originally installed at 111 E. Ontario. They were installed in a room near the front of the building.

I remember getting very frustrated trying to keep them up to "spec." I suggested to Bill at one time that we purchase new heads for the slave machines. He wasn't too enthusiastic, and suggested I spend more time in tweaking the system. They were still a bear to align. Since the slaves had no playback provision, I had to run the tape down the line of machines and through the playback machine. When I was doing the 3200 system, Ampex 350 machines didn't yet exist, as Jerry later talks about.

There was one echo chamber at Ontario, it was above Control Room A, past the tape library. I remember Bill taking me up there so that I knew about it. An odd room, plastered on all walls, ceiling and floor. A loudspeaker of some vintage, and I think the mike was a WE 639A. That was it! Almost nothing was parallel, as a long reverb without a pronounced initial reflection was wanted. We regularly kept silica gel (or the equivalent) suspended above a pail to keep the plaster very dry.

The loudspeaker system in the dubbing room was a Jensen G-610 Tri-axial in a curious wall cabinet. I never found out just what the cabinet design was, if any. For all of the curious cabinetry, that loudspeaker system was great for setting eq and levels. There wasn't a lot of equipment for equalization and level setting. I think there was a Cinema Engineering 4031B equalizer with an Altec-Lansing limiter, and probably a Cinema HP-LP filter. Nevertheless, the room was great for producing dubs in quantity. The turntables were very unusual, as a round belt drove each table from a Crocker-Wheeler motor, which rotated at a rather slow speed. When starting them, after applying power, each table had to be spun to bring it up to speed. The dubs were cut at both 45 rpm and 78 rpm, but 78's were quickly dropping out of favor. The dubbing system was rebuilt when we moved to Walton Place. Dale Harrington cut dubs in the evenings. I was told that he had been a reporter with the Chicago Sun-Times, but I never asked him if the comment was true.

Bill's wife, Belinda Richmond, taught me quite a bit about eq'ing on the fly. The two cutting rooms were later moved to the new studios on Walton Place. Belinda also taught me editing on the Soundcraft tape editing machines. They looked very much like 8mm film splicers, but without the locator pins and of course handled only 1/4 inch tape. Mind you, this was 1954, long before Joel Tall designed and sold his "Editall" splicing blocks. I learned from Belinda that she had been a singer before she met Bill, which explained part of the reason for her excellent editing ability.

Bill had become friends with Stan White (Jerry will talk more about him), who was a loudspeaker system designer and manufacturer. He brought in a loudspeaker system for Bill to try out. It was one of Stan's floor-loading cabinets, the top was about five feet, eight inches high, about 3 feet each in the other two dimensions, and used one (!) JBL 8" loudspeaker. A very impressive sound, indeed. I recall later that Bill and Stan mounted it in the control room wall, and loaded it to the room. Since stereo hadn't come along, speaker placement was at Bill's convenience.

I remember one session for Mercury with Ralph Marterie and his Orchestra. The band was usually on the road, so for the local sessions, Ralph used Frankie Rullo, a Chicago studio musician, who was a great drummer. In those days, the Musicians Union enforced a 3-hour session, with a maximum time of fifteen minutes of finished music. This night, there was about ten minutes left until the end of the session. Ralph called out the name of the tune they were going to play, and said it was going to be a jam session. He was standing in the middle of the studio, pointing to the musician who was going to do the next 16 bar solo.

Well, Frankie's style of playing when he was just doing backup rhythm was to put his head down and concentrate on the rhythm. He was a very stable drummer. At one point, Ralph spun around and pointed to Frankie at the back of the studio for his solo. Naturally, everybody else simmered down, the piano, bass player and guitarist started block chords. Unfortunately, Frankie wasn't aware that it was his solo, so he continued his rhythm beat. This kept on for about four bars, and then Ralph and the band broke up with laughter. At this, Frankie looked up to see what the laughter was all about. It was about him! That broke up the session.

One of the studio regulars was Earl Backus, who played electric guitar. His amp was an old-looking Fender, but it sounded very good. One night during a break, I asked Earl why his guitar sounded so good. He turned around the case and showed me the amplifier inside. It was a Williamson, well recognized in audio circles as being a great amplifier! He pointed out that he didn't show the amplifier to just anybody.

Bill wanted to cut some frequency run test lacquers for later pressing in short runs. He had an excellent announcer from the Chicago area, Ken Nordine, do the frequency announcements on tape. We then played the tape for each frequency statement. Ken Nordine had a magnificent voice; he later built a show that continues to this day called "Word Jazz." Anyway, he did the announcements, "1000 cycles, 500 cycles, 250 cycles, 100 cycles." He finished the run by saying (very close to the mike), "Zero cycles." Unfortunately, we didn't use it!

Bill then made an appointment to see Ben Bauer, who was with Shure Brothers (the microphone company), who also made excellent phonograph pickup cartridges. He wanted Ben to do light pattern (Buchmann-Meyer) measurements on the lacquers. Bill took me with him to meet with Ben Bauer. He was also very involved with microphone measurements. At that time, they used a WE 640AA microphone as the standard for their anechoic chamber. There I was properly introduced to light pattern measurements, which I later used, and still remember the technique.

We received some tapes for mastering from the West Coast. The artist was Gale Storm, who was signed with Dot Records. For her mike, the studio used one of the brand new U-47s. They were being imported by Telefunken, as Gotham Audio wasn't yet in business. Not to be too technical, but the European standard for mike inputs on consoles called for higher input impedances than used in the US. As a result, the U-47 was terminated at a lower impedance, and the top end was very edgy. We had to do some extensive EQ to make her voice sound normal. Later when Gotham Audio started importing U-47s, which were actually made by Neumann in Germany, Gotham began supplying translated documentation which caused many companies to take a new approach to microphone preamp design.

## 1955

**Bob:** Just like Jerry, I had no real idea of what I was doing. I learned a tremendous amount each night from Bill and Belinda. One night, Bill was doing a jazz session. I don't remember the musicians, except for the drummer, Jo Jones. Yep, the same Jo Jones you'll find on jazz LP's and CD collections. He was a great drummer, and would tend to make faces as he followed the piano player. Anyway, I was taking photos of the session (with Bill's and the producer's approval). I took several shots of Jo Jones from not too far away, and then went back to the control room after the take was finished.

After the playback, Bill glanced into the studio, and then quietly said to me, "Watch yourself, Jo doesn't look happy." Jo came in the control room, and neatly chewed me out for disturbing his playing. I apologized profusely; I don't think I took any more pictures of that session. It was then I realized the demands on jazz musicians, who are creating as they're playing. Once a phrase has been played, they can't go back and edit it. If they do, for many aficionados it's not jazz.

Sometime in 1955, Bill hired Tony Parri as controller. Tony was a short, mild-mannered Italian gentleman, who moved with Bill and Jerry to Los Angeles. Physically, Tony wasn't a very imposing individual, but as I later learned in Los Angeles, he had a keen understanding of finances. Bill's hiring of Tony was one of the best choices Bill ever made.

**Jerry:** Amen! I feel certain that Tony was indirectly responsible for my rise through the company.

I had a similar first meeting with Bill. I was in my senior year at Valparaiso Technical Institute, and was referred to Bill by the then Dean of Education, Cloid Patton. It seems that Bill had also attended Valparaiso Technical Institute a few years earlier and knew the staff. In particular, Dr. Hershman, the President and a well known acoustician later helped Bill with the acoustic measurements and treatments for the new Studio A at Universal on Walton Place.

**1953: Berwyn, IL**

**Jerry:** I had just graduated from High School in Cicero, IL and I had had already made a name for myself there as lead technician of the Visual Education Department. I was able to run anything they had (and they had quite an inventory) and to fix most of it. I had edited thousands of feet of 16mm Sound stripe film for the Superintendent of the district and added his narration and background music to it. I had worked for a commercial photo lab for about a year, took Drafting, Electric Shop, Machine shop and Welding. I didn't like math very much and didn't take any more than I needed to get into Junior College which happened to be in one wing of the High School building. After a year of Junior College, I decided I was wasting my time.

A friend of mine that I had met earlier in High School and who now lived in another village about 15 miles away was also interested in electronics and recording. We built our own binaural recorder using a bi-directional Webcor reel to reel tape recorder, by adding a second record/play amplifier to the existing extra heads in the Webcor. This gave us staggered ½ half track binaural recording. My friends parent's were well known at the local Country Club, that held dances on Saturday nights, and arranged for us to record a few. We didn't have much money to workwith but we managed to scrape up enough to buy two matching cardioid dynamic microphones. After tweaking the system, we were ready for the big time.

I went to the Director of the Music Department at the High School for whom I had done some filming earlier. He allowed us to set up and record a live public concert of the High School Symphony Orchestra in the Main Auditorium, a 1500 seat theater. The music department was also recording with the school's Magnecord recorder and a couple of Altec condenser microphones (mono only of course). When we showed up with our Webcor and cheap mikes they kind of looked down their noses at us. But at the end of the concert, when they came over to hear what we had recorded, they were flabbergasted. None of them had ever heard a true binaural recording, and they passed the headphones back and forth for half an hour.

At home, I had a tape recorder of my own; a Masco 1/4" full track machine. I had modified it to run at 7 1/2 ips (it came as 3 3/4 ips) and added an electronic erase head (the original one was a permanent magnet type). I also had a G.I. disk cutting turntable that came with a crystal cutting head which I replaced with a magnetic cutting head so I could drive it with the low impedance output of my Allied Radio amplifier. My Dad was a telephone engineer with Western Electric, and we had all the electric/electronic goodies to play with. Our whole house was wired for background music, and was controlled by a dial up system to turn on the amplifier and a couple of radios. So you can see where I got my interest in audio and electronics.

My friend's father had heard of a good electronics school in Indiana and my friend and I went with his father to look it over. We liked what we saw and I went to my Dad to ask if I could attend with my friend. He agreed, and that's how the story started. I attended Valparaiso Technical Institute for two and a half years, during which time I joined the Amateur Radio Club and was Chief Engineer of the campus carrier current radio station for a semester where I raised a few eyebrows by re-stringing the RCA Junior ribbon microphone with the foil from a chewing gum wrapper. The professors all began to realize how interested I was in broadcast and recording in particular.

Bill wrote me a letter asking me to drop in at the studio some evening when I was in Chicago to meet with him. When I arrived, he was in the middle of a session in the new Studio A and asked me to take a seat in the back of the control room until there was a break. During the break he showed me around and we talked. Apparently he liked what he heard and told me there would be a job waiting for me when I graduated. I was going to be a maintenance engineer and I was eager to get started.

**1956**

**Bob:** In early 1956, Little Richard did an R&B tune called "Long Tall Sally" for Specialty Records. Shortly after the record came out, Randy Wood, who had started Dot Records, came to the studio with Pat Boone and a 45-rpm record of "Long Tall Sally." I don't remember who wrote the arrangements, but the 45 record was played several times until Pat got a feeling for the tune. I remember Bill working with Randy Wood and Pat Boone doing cover records. That work Bill seemed to enjoy.

Bill had gotten a request from Decca Records in New York to record a performance of two of their artists, Josef Marais and Miranda. They were giving a performance at a small, but excellent hall in Chicago, located in the Art Institute. I used one microphone for them, fed to two Ampex 350 machines. They were very nice musicians, and were kind to a nervous engineer.

Later that year Bill asked me if I wanted to do another remote session. "Sure", I said, "why not" - but then I didn't know what I was getting into. Billy Taylor, who was (and is) one of the best jazz pianists in the U.S., was appearing at the London House, a good restaurant and bar. The London House was located near the intersection of Michigan Avenue and Wacker Drive, just across the river from the Wrigley Building.

Mind you, this was 1956, and real remote gear didn't exist except for some clumsy RCA remote mixer boxes designed for remote broadcast use. Ampex had just come out with their Model 600, which was a small 3-head recorder. We didn't have one, as we didn't do much remote work. So, I put an Ampex 350 (not a small machine) in a van, together with mikes and cables, and headed to the London House. I don't remember what I used for a mixer.

Billy played piano, Earl May on bass, and Ed Thigpen on drums. As I recall, the management tucked me into a small room with no view of the piano trio. It was very close to the patrons, so I had to use a pair of headphones. The neat headphones later developed by Koss didn't exist. These were Brush units (crystal transducers), good quality, but no bottom end.

I could hear the piano and drums, but I had to estimate the bass level by how it was kicking the VU meter. I met Billy before they started their first set, so he knew I was there. The producer showed up, and wanted to hear the mix on the headphones. I explained that he wouldn't be able to hear the bass, and about the way I was mixing. Being that he only had his ears, and not knowing about VU meters, nevertheless he had me turn up the string bass mike so he could hear the total mix. I tried to explain to him what we would get, but he decided that's the way it's going to be.

As I remember, it was early in the year, and rather cool outside. The producer wanted to get photos of Billy for the album cover, so they went outside. I tagged along, as I wanted to get to know Billy. Billy wasn't too happy at being outside, no coat, just his suit jacket. I could tell he was cold!

Finished for the evening, I packed up the gear, and went back to Universal at Walton Place. I had made arrangements with Billy to hear the tapes at Universal the next afternoon. He showed up, and then we went into the mastering room. Mix-down rooms didn't exist, yet. I started the tape with the first set, then we listened as the string bass level was brought up. The rest of the tapes, probably two sets, were string bass accompanied by piano and drums. Not a good mix! I tried doing some eq'ing, but the string bass was too high in level. So much for his London House session! I don't know if that was ever released on LP.

## CHICAGO, Walton Place
## 1957

**Bob:** I had stayed with Bill for about a year, and then decided (incorrectly) that R&B and the recording industry weren't for me. I worked as a foreign car mechanic, then at a record store in Evanston. I decided that the recording industry wasn't all that bad, so I called Bill about coming back to work for him. He had always treated me as an equal, not just the kid on the block. Bill had just leased the entire 2nd floor at 46 E. Walton Place, and was trying to supervise construction during the day, and do sessions at night. Mind you, this was my <u>second</u> time being hired by Bill. I was somewhat familiar with studio construction, so Bill thought I could do the job of construction coordinator.

Gypsum block was being used for the studio walls, and I recall showing the masons that the separate walls, although adjacent, couldn't be touching one another. As Jerry said, Dr. Hershman did the acoustical design, Curt Esser was the architect, but the timetable was so short, some of the design and construction was done on the spot. I'd see a problem, if I couldn't answer it, Bill would be contacted as to the answer. I don't remember all the various tasks I handled, but apparently Bill was satisfied.

The 2nd floor was to be office, mastering rooms, and workrooms. Since we needed high ceilings for the studio, Curt Esser worked with the contractor to redesign the area. Nowadays, the job would be called "fast-tracking."

We moved into Walton Place, and I stayed with Bill until around August 1957. My normal work was cutting lacquers. As Jerry well knows, after working in the studio, your ears (and brain) become educated. Even to this day, I'm very aware of how something sounds. High control room audio levels weren't the rule back then, so our ears stayed pretty sensitive. I recall that the limiter in the mastering room was manufactured by Altec-Lansing. It used a pair of variable-mu tubes for gain reduction. If the tubes weren't matched, which was fairly often, we had to select matching tubes. I think it was Bill or Bernie that showed me the technique of touching the grid caps to introduce a signal in both tubes to see if they were balanced. Otherwise, the limiter would "thump" when limiting occurred.

**Jerry:** We taped a lot of cover records in those days. These days, most artists or groups will only record their own music and rarely do covers. I remember staying up all night cutting seventy or eighty double sided dubs (on acetates) to be mailed out to DJ's early the next morning. Belinda would order a gigantic antipasto salad from the Italian restaurant in the basement and place the platter on the baby grand piano in Studio C. There we could snack on cold cuts and cheese and salad all night long as we worked to get out the dubs.

These were cut four simultaneously on a custom four-headed lathe designed and built by Bill for that very purpose (see pictures in the photo section). They were also used to cut advertising spots for multiple radio stations. We also pressed the two Scully lathes into service and had people typing labels and sticking them on the dubs and sliding them into their covers and into mailing envelopes for shipping. Everybody got into the act. Mind you, this was long before Compact Cassettes, much less CD's.

**Bob:** Just a short comment, I remember doing a couple of dub scenes at Ontario, but we didn't have an antipasto salad! (Jerry; One of the perks for having a studio built above a restaurant!)

**Jerry:** By the time I arrived on the scene at Walton Place, there were also a number of Ampex 350's and a whole slew of Ampex 3200 high speed (60 IPS) duplicator machines (based on the 300 deck) for their pre-recorded tape line, Moods in Music... mostly jazz.

Bill certainly did relate to the clients and musicians! He was on a first name basis with all of them and was highly respected as a mixer. The list of his clients was long and the names legendary. Who were the clients? Len and Phil Chess of Chess Records; Art Talmadge and Nook Schrier, (Nook's professional name was David Carroll) of Mercury Records; Randy Wood of Dot Records, the Ertegun brothers, Nesuhi and Ahmet at Atlantic Records to name a few. Who were the artists? Muddy Waters, David Carroll, Ralph Marterie, The Crew Cuts, Pat Boone, Billy Vaughan, Les Paul, Herman Clebanoff, Remo Biondi, Peggy Lee, Duke Ellington, Stan Kenton, Count Basie and many lesser-known R&B artists from the Chicago area.

The new dubbing turntables at Walton Place were driven by a single motor driving a 1 1/2-inch wide cloth belt that traveled around most of the circumference of each turntable using a series of roller guides that supported the belt between the turntables and back to the motor. The motor had a stepped, pulley that allowed changing speeds. This assured that all tables were turning at the same speed, and isolated the tables from motor rumble. Each table rotated on a chrome ball in the shaft well, and weighed about ten pounds. The belt would slip a little on the crowned motor pulley getting up to speed due to the huge mass it was trying to accelerate.

**Jerry:** Acoustic echo chambers were catching on. Universal on Walton Place had two echo chambers, one on the roof for Studio A, and another behind the mastering room as a utility chamber for Studio B (as yet uncompleted) and Studio C. At Walton Place, Studio C was very small, and had a converted Western Electric console in it. (Bob; that was the board you and Bill had modified at 111 E. Ontario.) It was used to record spots (commercials) and jingles during the day and lots of R&B at night.

**Bob:** Incidentally, all the faders were rotary (slide faders either hadn't been invented, or hadn't made their way here from England). On Bill's boards, there were no dials around the knobs. He didn't need them. Doing some mixing later on, I found he was quite correct. Bill's two partners, Bernie Clapper and Bob Weber, worked during the day on agency (advertising) sessions. Bernie would ask me almost every time I saw him if I could put fader dials on the board. I always ducked his question by saying, "I'll ask Bill where they are." Bernie would regularly put grease pencil marks around the faders to remember level settings. For agency work, the balance was more critical - at least, that's the way Bernie put it.

**Jerry:** Having done some film dubbing later in life, I know now what he meant. Bill used to tell me to clean off all the pencil marks whenever I checked out a studio. Bill's consoles were custom made by Universal Audio, a subsidiary of Universal Recording in Palatine, a western suburb of Chicago run by a gentleman by the name of Roy Rogers. No, not the cowboy, though he did have an old Studebaker he called Trigger. Roy and Bill designed the consoles, and Roy and a couple of helpers fabricated them there in his shop.

**Bob:** The board in Studio A on Ontario was probably one of the first custom boards from Universal Audio.

**Jerry:** I believe most of the previous consoles were modified commercial broadcast types, but the new Studio A on Walton Street had a completely new one, built from scratch especially for recording. It had echo send and receive pots and three output busses. There was a left, a right, and a monaural buss. Since the control room was set up for monaural listening, a switch let you listen to any ***one*** of the three. There was a large Stan White designed speaker system on the right hand wall of the control room, and a Mcintosh amplifier to drive it. In those days, 75 watts was a lot of amplifier, often weighing in at 25 pounds or more due to the weight of the chassis and transformers.

I'll never forget the time shortly after I started working there that Bill called me in and said that the power supply on the console (actually two supplies with a changeover switch) was getting noisy, and was causing pops and snaps in the console output. He asked me to pull it out Friday night and fix it. I pulled it out and took it to our maintenance lab for repair and found it needed some new electrolytic capacitors that were arcing internally (in those days, it was not uncommon to have four or five hundred volts across the filter capacitors in console supplies). I pulled out the capacitors, wrote down the numbers and figured I would pick up new ones on the way to work on Monday. I was totally naïve about the operation and didn't realize they sometimes worked Saturdays and Sundays at recording studios. You can probably guess what happened...

Bill came in Saturday afternoon to set the studio up for a session that evening and saw the power supply was missing. In a panic, he pulled a supply from another studio, stuck it next to the console and hooked it up with clip leads. Then he went back to setting up the studio and did his session that night. I heard the news when I arrived Monday afternoon (I worked nights then). Boy, was Bill steamed up over that! However, he cut me some slack because I was new, and I was very careful not to put anything out of service without checking with the office from then on. About that time I also made up some Maintenance Request forms so that I could keep on top of repairs. Those consoles had over 40 tubes in them not counting speaker amplifiers and power supplies.

While we are at it, we probably should describe Studio A at Walton a bit more. The studio was about 40 feet wide and about 75 feet deep. It was big enough to hold a full symphony orchestra, and on occasion actually did! The ceiling was about 20 feet high and was treated with acoustic tile that had 1/8 inch holes spaced about 1 inch apart. There were 8-foot fluorescent fixtures hung from the ceiling. I don't remember if the ballasts were remote or not, but I don't think so. They were high enough so that very little noise was picked up below. As I recall, the air conditioning was louder than the noise from the lights.

The walls were built on the staggered stud design, and covered with Masonite pegboard over fiberglass insulation installed with the fuzzy side toward the studio. On the walls in strategic places were hung 10-foot high panels suspended at one edge by huge piano hinges. These panels were poly-cylindrical on one side, and the afore-mentioned pegboard/ insulation treatment on the other. These could be swung about to change the acoustics of selected areas of the studio. This was way before the idea of "live end, dead end" studios became popular. The walls were isolated from the ceiling and floor using rigid foam rubber.

The floor was rubber tile over a concrete slab that was isolated from the rest of the building by a slot around the perimeter to prevent impact sounds such as drums, etc. from being transmitted to the hallway or to adjacent Studio B. The two studios used similar construction techniques and were separated by a hallway with a freight elevator at the end (the studios were on the second floor, after all). The elevator went all the way to the basement, where there was an Italian restaurant (a favorite hangout of Bill and Belinda's). The enticing smells of the restaurant kitchen frequently wafted up the elevator shaft and into the back hallway. So also did smoke when the restaurant had a minor kitchen fire. That one had firemen running up and down our hallways when a client called the Fire Department after seeing smoke coming out of the elevator shaft. The elevator was also used to bring up the larger instruments from cartage trucks in the rear alleyway. Instruments such as pianos, percussion sets such as tympani, chimes, gongs, etc. were routinely moved in and out for musicians coming to sessions.

The control room was about 12 feet deep and 18 feet wide. It was located about 10 feet above the studio floor along the long side (West) of the studio near the main studio door. You entered it by going up a short flight of stairs from the hallway. All the doors were of the standard acoustic type in used in those days (a solid outer frame with center section suspended in foam rubber). Bill would use his own design when he later built the studios in Hollywood.

**Jerry**: When I arrived on the scene at Walton Place, Studio B had not yet been completed. The console, which was pretty much a twin of the Studio A console was not yet finished and had not been moved from the Universal Audio shop. One of my first jobs was to hang the fluorescent light fixtures on the ceiling and get them hooked up to power. I had to lie on my back on top of a scaffold to install them. I remember that the holes in the ceiling panels (about three feet from my face) tended to give me vertigo until I learned not to focus on them. I also had to pull the wires for the microphone connectors and the cue circuits from the connector box into the control room.

Bill hired a couple of telephone guys to install the wires on a telephone "tree block." He said that they could do it faster and neater than anyone else since they did it all day for a living. (I know I wasn't good at it). He was probably right and I later did the same thing on other projects when I got to Los Angeles. By the time the console was delivered, we had the rest of the wiring in place and it was just a matter of connecting everything together. I didn't know it then, but I was to repeat that job many times during my career in studio work. Bill liked to use three-wire Hubbell "Twist-Lok" connectors for microphone connectors because as he said, there was always a clean connection due to the wiping action when you twisted them, and they didn't fall out easily if someone tripped over the cable. This worked OK where you weren't swapping microphones with other studios or renting microphones from other sources.

Because Bill had contacts with the phone company engineers, he was able to get telephone cable at Graybar, a wholesale supplier of all things telephone. Most of the work areas at Universal were tied together (electronically) by multi pair unshielded telephone cables. These worked fine for line level signals most of the time and there were no real cross-talk problems with monaural signals. I don't know who installed these cables, telephone guys or electricians, as it was well before I got there.

One of my first big projects was modifying one of our older Ampex 300's to get rid of the horribly noisy 12SJ7 playback input stage. Ampex must have known it was noisy since they had provided a weighted, padded metal cover over the tube. While this worked to some extent, the tube was a metal type and with the cover on, it got quite hot which caused it to get noisier and shorten its life considerably. We ran the machines without the cover most of the time. Bill and I tinkered around with the circuit for quite a while until we had a satisfactory noise figure. Tape was noisy enough those days without adding more in playback, especially if you were going to re-record it again for an over-dub.

The next big project Bill handed me was converting an Ampex 300 deck to a 3200, as we needed another high-speed duplicator slave unit. This entailed modifying the transport controls and replacing the capstan assembly for one that ran at 60 I.P.S. I took the oldest 300 we had out of service as it was not used very much since we had several new Ampex 350 machines. It worked as well as the rest of the group (we had six slaves and one master playback unit). These became part of my nightly routine, as they constantly needed tweaking.

Aligning these machines was really a pain. You ran the tape down the line of slaves and through the playback machine. Each adjustment took a couple of seconds to register on the meters. I made up some adapter cables that enabled me to use the slave record heads for playback so that I could align the heads on the slaves by playing the alignment tape on the record heads, using a spare Ampex 350 playback electronics that I just moved from one slave to the next. It cut down the time it took to align the slaves by at least two thirds. This was a boon since we were aligning these machines every night after the operators went home.

**Jerry:** I also had to do the routine maintenance on the Mastering Room and the Dubbing room about once a week. Bill taught me how to run the Scully lathes and adjust the cutting levels. I would check out the mastering console (a small table mounted rack containing equalizers, filters and a limiter. There was also some specialized equipment related to the cutting of disks used to provide sound for filmstrips in educational settings. I knew a little about this as I had used these many times as a Visual Education Operator in high school. Bill's partners, Bernie and Bob did quite a bit of this kind of mastering for their clients.

The lathes had Cook Cutting heads mounted on them and Bill taught me how to tweak the frequency response of these heads by cutting notches in or gluing an extra thickness on the "Viscoloid" material that surrounded the armature and provided dampening for the stylus. The control for the feedback also had an effect. The power amplifiers were Presto units that Bill had modified to accept the feedback from the Cook heads. Bill bought a Neumann Collimated Light Source to check frequency response by using light patterns. Boy, I had no idea when I applied for the job that the learning curve would be so steep. Just learning to cut and read the light patterns was an exciting new process for me. But it sure made adjusting the response easier and much faster as I could make adjustments on the fly.

One night Bill said that he had ordered a tuning fork generator for the special unit in the mastering room and asked me to install it when it arrived. It came on a Friday morning, and since the mastering room was not using the unit at the time I removed the unit from the rack during a break in the afternoon. The system was made by the DuKane Professional Equipment Company (the company that made the special filmstrip projectors for schools and businesses). It sharply filtered out the very low frequency audio of the program material and added a low level 50 Hz tone to the audio. This tone was used to lock out the projector mechanism so that stray audio would not advance the film. When it was time to advance the film, it dropped the 50 Hz tone and inserted a 30 Hz tone that activated the film advance. The system was referred to as a 30/50 system. The frequency and timing of the tones was very critical. The original unit used a tube oscillator to generate the 50 Hz tone, and a tuning fork oscillator to generate the critical 30 Hz tone. There were no phase locked loops or digital filters in those days. Everything was done with discrete or electromechanical components. The fork Bill had ordered was a 50Hz fork so that we could upgrade the unit to be the same as the newer ones.

I had to modify the chassis to add another tube to drive the fork. I had to make some room on the chassis to mount the fork unit, a rather bulky package about 2 inches wide by 6 inches long. The existing fork was also that size, so they took up a lot of real estate on the chassis. These were not miniature tubes either, so the whole unit was quite large and heavy. They had booked the system for a client on Monday morning at 10 o'clock, so it had to be built, tested (we had a record player/ filmstrip unit to test the system) and installed before the client arrived. I had some problems getting the fork to oscillate, and had to change some of the circuit values to get it to start by itself and that slowed me down. I worked from Friday afternoon to 9 o'clock Monday morning straight through with only a couple of 10 minute catnaps. It usually took about 15 or 20 minutes for the forks to stabilize to their operating levels so by the time I had cut and tested a quick dub of the system, the first of the clients were arriving. I was so exhausted, I had to take a quick nap in my car before driving home whereupon I slept for 16 hours straight before I felt refreshed. That happened on only one other occasion since then (in Los Angeles).

I remember when Dick Tullis built the enclosure for the Stan White speaker from Ontario into the wall of the control room at 46 E. Walton Place, sticking out into the sound lock. Of course, it had to be an insulated wall to prevent spilling audio into the sound lock, so it was a double- studded isolated wall of the type Bill used in the studios. Bill had the idea of filling the inner space with sand to increase the mass of the wall so it wouldn't transmit bass into the lock. They got several bags of sand at a construction materials yard, and filled up the wall through a hole in the sheetrock near the top. All was well for a couple of days, until Bernie came in one morning to a sound lock full of sand. It seems the sand was damp, and the trapped moisture weakened the sheetrock which collapsed under the pressure of several hundred pounds of sand.

Stan (White) was a frequent visitor at the Walton Street Studio. He was always bringing his latest speaker system in for Bill to listen to and critique. Bill would bring out one of his favorite tapes to play through the speaker, which they then compared with other speakers in our studio. Sometimes this lasted for hours.

I remember one particular time Stan brought his latest speaker design for Bill to check out. It wasn't mounted in a cabinet or box; it just sat there for all to see. It was a 15 inch woofer (in those days they weren't called sub-woofers) and the cone was made of solid molded polystyrene plastic. The kind coffee cups are made of, but much thicker. It had a special shape (cross section) that he had designed mathematically to avoid cone breakup. The big difference in this speaker was the 15 inch diameter voice coil that was made of edgewise wound copper ribbon embedded in the outer circumference of the plastic cone. The cone was suspended in a magnetic field created by several dozen flat magnets of the type used in cabinet door latches. They were cemented to the inside of a circular cast aluminum frame. This made for a very lightweight unit compared to the same size unit utilizing the standard magnet and frame of then current woofers. The cone being solid, but very lightweight foam plastic was a perfect piston driver and although only a prototype performed remarkably well. The cone moved as one solid piece from edge to center and could move about 1/2 of an inch on either side of center. It moved a lot of air! Bill had Dick Tullis (who had once worked for Stan) build a bass reflex box to mount the speaker in, and we had BASS! Stan said at the time that it would be even better in a horn enclosure that was the type of enclosure he preferred and used in most of his systems. I don't have any idea what became of that prototype, but I have never seen it built commercially. I was very impressed with the idea then.

We tried many different types of speaker cabinets through the years. Most were discarded because Bill didn't like the sound of them or he'd heard something he thought was better. Somehow we always came back to the basic bass reflex cabinet, usually with some modifications.

Stan White frequently brought various speaker systems for Bill to evaluate, as he valued Bill's opinion. Stan approached building speaker cabinets from a theoretical standpoint, and his cabinets were mathematically designed. He favored cabinets that featured a tuned horn for bass reproduction. Since space was at a premium, most of the horns were folded types. He also had developed a formula for an exponential horn that, while enclosed in a relatively small box, had a very smooth fold to avoid reflections in the horn throat and they loaded to the floor and the wall behind the cabinet (they stood on legs) which formed the mouth of the horn. I thought they had amazing sound, as he was using relatively inexpensive 8" speakers with a 5" paper cone tweeter. I still have one of those speakers to this day. Stan had also developed an amplified high frequency driver utilizing a specially built four inch stiff paper cone with dual voice coils. One drove the speaker and the other was used to provide feedback to the amplifier to reduce distortion. He also had a couple of very large (theater size) systems that were truly astounding. They were four feet wide by eight feet high, and contained two 15" bass drivers in the bottom horn section and two midrange drivers coupled to exponential horns in the center section plus an aluminum domed high frequency driver, also coupled to a horn at the top of the cabinet. These could easily provide sound to a very large theater, and were surprisingly efficient as they could be driven by a 100 Watt amplifier.

The speaker in the mastering room at Walton Place was a Stan White corner horn experimental design with one 15 inch, one eight inch and one 3 inch metallic driver. It sat on two 8 inch brass legs in front and a "V" shaped leg in the back. The finish was blonde oak, a popular color of the day. It stood about five feet tall and was about thirty inches wide. This speaker was one of my favorites as it had a very smooth frequency response and very nice, solid bass response. There were actually two horns built into the speaker box, one for the bass of course, but also one for the midrange speaker.

The floor playback speakers in the studios were stock Altec A7 "Voice Of The Theater" systems.

The speaker in the dub room was a 12 inch Jensen triaxial unit in an infinite baffle mounted in a corner above the equipment rack.

**1957 (later)**

**Bob**: I decided that a college education would be good for me, so I left Bill and went to the University of Illinois in Champaign. Bill never said anything, but I suspect he was disappointed when I left. When Jerry and I later met at United Recording in Los Angeles, we figured out that we had just missed one another by a month or two in Chicago.

After one semester, I realized that a college degree wouldn't do much good for me. I lived with my parents for a couple of months, figuring out my choices. The recording business was the best choice! I moved to New York and went to work for Tommy Dowd at Atlantic Records. I became reacquainted with Billy Taylor when his trio was playing at a jazz club on 52nd Street; I dropped in at least once a week.

Tom had Atlantic buy the third Ampex 8-track tape recorder ever manufactured. (The first two went to Les Paul). The studio was on the fourth floor at 234 W. 56 Street, and the stairwell was too narrow for the moving people to carry the racks. They ended up by suspending each rack below the elevator! The electronics were in one rack, and the 300 transport in another. The channel to adjacent channel cross talk was only about 30 dB. Remember, this was 1957!

Atlantic's mastering system was a Scully lathe, with a Cook feedback head driven by a modified Langevin amplifier. I still have some of the mono LP's I cut on that system. Cutting lacquers at Atlantic made me realize how thorough my education had been at Universal Recording!

I met Larry Scully, who came to our studio in response to a maintenance call from Tommy. From Larry, I got some valuable pointers on maintenance of Scully lathes. That learning was of great help in Los Angeles and Mexico City.

Tommy used a 45 rpm pressing titled "Boll Weevil," sung by Teresa Brewer on the Coral label as a benchmark. If we could get something we were mastering to sound as loud as her record, we were doing good! In those days, louder was better because when the record was played on a fixed level jukebox in some ice cream parlor or tavern, your record stood out from the rest of the bunch and was likely to get more play.

I remember Tommy and I looking at an LP pressing of the first stereo record made. The lacquer was cut by Westrex, and surreptitiously pressed by Audio Fidelity. Thinking at the time, wow, is this going to change the state of mastering!

**Jerry:** Belinda taught me to edit tape. We used funny looking machines made by Soundcraft that looked similar to film splicing machines. I was very familiar with those from my days in high school, so I took to the idea very quickly. I had had some musical training when I was much younger, so I had a feel for rhythm, counting bars, etc. This all fell into place one day when Belinda was under the weather and I had to take her place in an editing session with Nook Schrier (David

Carroll) from Mercury Records. I had never edited with a client before and I was very nervous.

We were editing an album for the Platters that had been recorded in Paris. The tapes had been recorded on hubs as was the practice at some studios, so they were shipped without reels to save weight. They were very tricky to maneuver during rewind and fast-forward and I was very cautious about it. We had made a metal base for the tape out of a discarded lacquer blank and used a flange from a 10 ½" tape reel for a cover. This was all held together by an Ampex clamping reel knob. The tapes were all numbered by the U.S. Customs Department and had to be returned to France in a relatively short time, or Mercury would have to pay import duty on them. Of course we could not cut the tapes, so we had to copy the sections we needed to a second machine for assembly. The music was very lush and full of strings and horns and was recorded in a fairly live environment. Nook walked me through the whole thing, telling me where he wanted me to make the cuts. Sometimes I couldn't hear a beat because of all the reverberation hanging over the cut. That was when Nook taught me how to keep time up to the cut and stop the tape on a mental count instead of what I was hearing. It worked, and I have used that technique many times since then.

These tapes were stereo originals, and some of the tunes had as many as eight or ten splices. We became very good friends, and he started asking for me to work with him. I think Bill had something to do with that as he told me I was working out O.K. If you knew Bill, that was pretty high praise. Bill was not particularly easy with pats on the back, but he was fairly quick to point out when he was displeased. We got along just fine.

I too was recruited to do some remotes at local clubs. Mr. Kelly's on Rush Street was the scene of one recording I did with emerging star Sarah Vaughn and her accompanying trio. Another time it was the Down Under Club for a recording of comedian Lenny Bruce to be used as research material for a writer doing a piece for Playboy Magazine. (The Playboy Mansion was just down the street from the studio).

Near the end of 1957, some of our ad clients talked Bernie into doing some recording for their TV ads. This entailed recording on the set while the film was being shot. It meant we had to rent equipment to do this work, as we had no mike booms or shotgun mikes such as are used on sound stages. Also we didn't have any portable tape equipment, or any way to record camera sync on the tape. We rented the equipment a few times, and then Bernie and Tony Parri (our comptroller) decided that we needed our own equipment.

We bought a couple of fish-pole booms and a couple of mikes and since Universal had bought most of their Ampex machines as portables (then made their own console cabinets), I remounted a 350 back in its portable cases and added a pilot tone head to the machines deck to record the sync. This stuff was all Greek to me at the time, but I read up on it and was up to speed pretty quickly. Bernie began taking me to the sound stages with him to help lug the heavy gear and to act as recordist.

A couple of times we also had to rent 16mm sound recorders and run them on a special cable which made them run with the cameras. More Greek! More boning up! I had studied synchros in tech school and was fascinated by the way they worked. When I found out this was the way these camera/sound recorders worked I was excited to learn more. About this time a gentleman by the name of Mason Coppinger came to work at Universal. He was familiar with the ad agency people and was to help out with the film business.

In a short time, it was decided that we should have our own film dubbing studio. We had the space, as a fairly large portion of our floor space had not yet been developed (we had the entire 2nd floor in the building on the NE corner of Walton and Rush Streets). Since I was still considered a maintenance engineer, I was asked to help build the film studio. Bill also recruited an engineer from a film company in town who arranged to buy 35mm and 16mm film transports for our new studio. I believe his name was Frank Richter. The machines were RCA film transports just like the sound heads that are on theater projectors. They were magnetic playback only, having been converted from optical units. I can't remember what brand the recorder was, but I think it might have been Stancil-Hoffman. I do remember that it could be converted from 16 mm to 35mm by changing the sprockets and plug-in heads.

Bill then asked another young man (who had been working for Stan White assembling speakers) if he would come and help with the construction of the studio rooms, as he had helped to build the other studios earlier. His name was Dick Tullis and he and I became good friends. Bill also recruited Larry Kissner who was working at Universal Audio to come and help with the maintenance and building. Larry and I worked together to install electrical conduit and wiring to all of the film transports, the projectors and the audio mixing console which was being built out at Universal Audio in Palatine.

**Jerry**: We were installing the three-phase motor/generator set that drove the projectors and film transports in sync with each other when I decided that the best source of three-phase power would be in the elevator control room which backed up to our new maintenance shop where we were installing the genset (called a distributor in film lingo). It was placed there because it kept the motor noise from being heard in the studio across the hallway. I drilled a hole through the back of the power panel on the rear wall of the elevator room, and we put a piece of conduit through the hole and I was fastening it with a nut inside the panel when my screwdriver slipped off the nut and struck one of the power busses. The 240 volts blew off the tip of my screwdriver and vaporized it throwing hot metal back at me. It was fortunate that at that time I was wearing glasses that kept the metal from hitting my eyes, but it along with the jerk reaction did tear up my right hand a bit and I was treated at the emergency hospital for shock and had six stitches in two of my fingers. I went back and finished the wiring before I went home so I could put the elevator back in service. Quite a day!

Early on, we started to record two-track stereo using a recordist to run the stereo tape machines in another room. This was a departure from the way earlier stereo recordings were made and originally was done as an experiment. Older stereo recordings had been made specifically for the market still in its infancy, and were recorded on a Magnecord PT6J-2. This machine had staggered half-track heads, necessary since the heads in those days were not designed for low cross talk having been made for monaural use. By the time I arrived on the scene, Ampex had in-line heads available with 30 to 40 dB of cross-talk rejection. Still not as good as staggered heads, but now tapes were editable and interchangeable on any machine with inline heads.

Bill had me move the shop to the rear of the building and prepare the vacated space for a stereo mixing room. We hung speakers on the wall (at that time they were Ampex 620 amplified speakers), and I built a small six channel mixer console with three channels left, and three channels right. It had two program output busses with meters and coupled master control with trims for fading. This served as a remote mixing position that was subservient to the main console in the studio. The idea was quite simple, but very effective. Particularly so in those days, when we had to exaggerate the separation effect to compensate for the poor channel separation in the early stereo cutting heads. It worked like this: The engineer in the main studio mixed his monaural product as usual, but split off the channels as left and right, combining them downstream as his monaural mix. The left and right channels were fed to the back room via tie lines. So far, so good.

Now comes the tricky part. Additional microphones were placed where they could pick up an overall picture of a particular section i.e. strings, percussion, brass, etc. These were fed directly to the back room via a couple of outboard preamps and tie lines. They were fed into the console on the opposite side of the console from that of their pre-mixed counterparts. Just enough of the sound from these microphones was used to enrich the stereo ambiance (a procedure that could be very tricky) so as to get the desired effect without overdoing it.

The idea caught on, and we had visitors from many studios around the country, who came to study the idea for their own use. The first real test of the system was for an album by Herman Clebanoff of the Clebanoff Strings on Mercury records. It debuted at the Chicago Hi-Fi Show at the Palmer House Hotel and was quite a hit. Incidentally that was the year that Stan White was criticized because his bass was shaking the second floor and rattling windows in the Michigan Avenue hotel. Stan had placed twenty of his smaller Opus speaker systems at three foot intervals around the walls of the room, and was playing a popular Hi-Fi demonstration record of the day, "Echos of the Storm", with a high power (for that day) 300 watt amplifier. All speakers were in phase and pointed toward the center of the room. It could be heard on the floor below.

One of my favorite recollections of working at Universal was doing a remote recording of the Indianapolis 500 at the Indianapolis Speedway for Mercury Records. We put together an impressive array of mikes and booms and all sorts of mounting hardware. We rented a truck and set it up with five stereo and one monaural Ampex 350 machines, each with its own monitor speaker. The monaural recorder ran continuously at 7 ½ IPS recording the track announcer's commentary during the race.

There was a machine assigned to each corner of the track, and one was switchable to cover the backstretch or the winner's circle. Bill and I rigged up pairs of mike pre-amps with big A and B batteries that we turned on just before the race. The machines were started at five-minute intervals just before the start of the race, and each time a tape was changed, the time was logged so we knew where each interesting sound could be found. Anything exciting was logged by one of the volunteers from the Studio.

The track telephone company (they had their own) ran audio pairs from each of the sets of microphone pairs out on the track to the location where we had our truck parked next to the grandstand. I got a large surprise when I started to hook up the pairs coming into the truck. There was 100 volts or more floating on the lines coming from the track mikes. It was enough to create sparks and give me a jolt. I didn't dare connect these to the tape machines for fear of blowing something up. I showed this to Bill, and he said, "Uh, oh, you need some repeat coils."

My dad was a telephone engineer, so I knew what a repeat coil was, but I had no idea where to find some, and I needed them fast as the race was the next day. I called the telephone guys, and they said that the high voltage on our pairs was inductive from running down underground raceways parallel with track power lines, and they had assumed that I would know all about that. I didn't, of course, or we would have brought coils with us. Another lesson learned.

The phone men didn't have enough coils to cover us, but Bill knew a guy in Chicago who worked for the phone company and was able to borrow 12 coils for a few days. One of the people from the studio brought the coils with them when they came from Chicago the morning of the race, and I hooked them up as we turned on and checked out all the mics before the race. Bill drove a borrowed golf cart all the way around the track keeping in contact with me by walkie-talkie (also borrowed from Motorola in Chicago). It was nice to be so well connected and have so many friends in high places.

Everyone from the studio came to the races including some people from Mercury who brought a big banner for the side of the truck exclaiming they were recording the "500." I still have that banner to this day! Everyone had a great time, and we got some great recordings. Freddie Agabation, one of the drivers that day, later did the narration for the record in the studio. When we finished, he gave me an autographed postcard of his car. I remember there was a place during the race that a car spun out coming out of a turn on the backstretch and hit the retaining wall. It was halfway between our corner microphone setups and so it was very difficult to make out with all the other motor and tire noises going on. We had the track announcer tape to use as a spotting guide, and if you knew what to listen for, you could pick out the crash. Someone at Mercury decided that we needed to "sweeten" the track "a bit" to bring out the crash. So Bill and I and another engineer (I'm not sure, but I think it was Malcolm Chisholm) set up a crash site out in Palatine on a rarely used back road. We set up stereo microphones and a portable recorder and had a tow company drop a scrapped car from as high as they could hoist it onto the concrete road. Then they dragged it along the road in front of the microphones for about fifty feet. The end result was that the listener heard a bump on the right and a sliding screeching going across to the left ending in a crash on the left. Mixed in with the actual track sounds, it was most convincing. That was my first experience with "doctoring" a sound track. I was to do it again a number of times in the future but never as blatantly.

There were many other times that I was called on to find sound effects for clients. Once, (a few years later) in San Francisco while I was filling in as Studio Manager and bottle washer for Don Geis who was on vacation, I was doing an advertising session for a regular client whose radio spot required a number of sound effects. These were to be the background sounds of a cocktail party going on under the whispering voice of the announcer extolling the virtues of the sponsors paté-in-a-squeezable tube for parties. After they all left, I couldn't help making a version for fun, and I substituted the sounds of a bar fight for the cocktail party and ended up with a police car with siren screeching to a stop and the sound of running footsteps and a door being smashed in. All this under the quiet voice of the announcer, I thought it was hilarious, and so did the producer whom I later played it for. However, he didn't think the sponsor would appreciate it.

## 1958

Sometime in June of 1958, Bill came to me and told me he was going to sell his interest in Universal to the other partners, and move to Los Angeles to start a new studio. I had heard rumors to that effect, and Bill had been making numerous trips to the Coast, but until then, nothing had been said about his move. I didn't think anything about the trips as Bill did many jobs as an acoustic consultant for other studios all over the country. I was dumbfounded and pleased when about a month later, Bill asked me if I would like to join his team in Hollywood. Not having ever been west of Illinois, I didn't have the slightest idea of what that would be like, but I was always up for new experiences, and couldn't wait to tell my wife Louanne. She was expecting our first child at that time and said, "What about the baby, and what about our parents who would want to see the baby?" I went back and talked to Bill who agreed that we should stay until after the baby was born and able to travel. He also agreed to pay our moving expenses, so we decided to open a new page in my career, and in the story of our lives. We had no idea!

**LOS ANGELES**
**1958 (December)**

**Jerry:** After settling into an apartment about 20 miles from Hollywood, I began to work at what was now named United Recording Corporation (Universal already existed here) on Sunset Boulevard near Gower Street in downtown Hollywood, California. The studio was being built in a two-story building that was once a film studio located just north of Columbia Pictures. They used to film short subjects about animals, and the place was full of animal stalls. Our workshop and office spaces were on the second floor, and we shared the building with another company called TV Recorders, an established film dubbing company owned and operated by the Aicholtz Family (a wonderfully friendly group).

During the day, Bill and I, and the people from Universal Audio (whom Bill had brought to Hollywood to build the consoles for the new studios), plus another young Hungarian emigrant (Balasz Nagy) worked in the upstairs shop building consoles. Bill also kept an eye on the workmen downstairs that were tearing out the animal stalls and building the studio rooms.

At night after the workmen had left, I worked a few hours pulling cable and wiring between the rooms through conduit placed by the electricians during the day. Like the studios in Chicago, there were wires that connected all of the rooms together. Unlike Chicago, however, the wires didn't loop from room to room. I got the idea of running the wires to a central point and creating a patch bay where any room could be connected to any other room via patch cords like a switchboard. I also decided that it would be better to run separate shielded wires instead of unshielded multi-pair telephone cable as some of the runs were quite long and we were running some low level signals at times.

Since we were installing several reverberation chambers, I felt that we should also put the cross-connects for them in that central patch bay. Bill agreed and so did Bunny (Robyn), our de-facto chief engineer and the original owner of Master Recorders, which Bill had purchased early on to use as a base of operations while our studios were being built. I bought the cable from a local Belden distributor by the caseload and strung the 1000-foot spools on a long piece of conduit supported on sawhorses. I pulled as many as 18 individual pairs at a time through the 1 1/2 inch conduits from the various workrooms to the central rack. In those days, there wasn't any multiple pair shielded cable available that was flexible enough to pull through the conduits. It was quite a chore and we had to lubricate the bundles of wires with special grease, but in the long run it was all worth it. I didn't see too much of my wife and new baby during that time, but we had most of our weekends together.

I got to design and build the power supplies for the consoles, and decided to use solid-state rectifiers to help keep the heat down. In those days it was necessary to use two rectifiers in series to provide enough reverse voltage rating. Also, since these supplies had to operate 30 tube amplifiers, they had to supply considerable amperage. Bill's famous cascode type pre-amplifiers were used, and the program amplifiers would put out a clean watt of power into a 600-ohm load. We built two of these consoles right there on the premises. The one for Studio A was finished first, and when the studio was ready, we slid it down the stairs to the lobby and around through the sound lock doors into the control room.

This was a 3-channel console and it used another of Bill's innovations. He had devised a way to fade all three master pots by linking them with belts and pulleys to a fourth knob which had felt clutch pads between the three pulleys it turned. This allowed the user to push down on the spring-loaded knob to lock all pulleys together for fading and still allow any one of the pots to "bottom out" while the rest were still turning. Great for fading the ends of tunes which was so popular in those days.

The power supplies for the console were two diode-isolated supplies, each capable of supplying enough power to run the console, giving us redundancy in case of a failure. The filaments were run on DC to keep hum and noise to a minimum. We used Electro full wave filtered power supplies capable of 15 Amps at 12 Volts. The power amplifiers were 60 watt, Universal Audio/Dynaco. I remember there was a session scheduled for the next morning, so we had to have it all hooked up and tested by 9:00 AM.

About 11:00 that night, the power went off in our part of Hollywood. We had flashlights, but no way to heat our soldering irons. We turned on the battery radio to see if there was any news of the blackout, and KMPC was on standby generators at their studio just down the street. The turntables had a bad "wow" from the generators, and the DJ kept apologizing for it, on a mike with bad hum. We all got a laugh from that, and sat around telling studio stories until the power came on 3 hours later. I'll never forget finishing soldering wires while the musicians were arriving for the session the next morning! Everything went smoothly, and everyone relaxed. I went home to sleep!

At this point I think we should say a few things about the studios themselves. Studio A, the largest of the three, was constructed using two entirely separate walls. The inner studio wall was saw-tooth shaped to eliminate parallel surfaces and was supported by but did not directly touch the outer wall, which supported the floating ceiling. Neither wall touched the floor, being isolated by various materials such as Celotex and foam rubber machinery mounting material. In effect a room within a room where none of the inside room touched the outer room directly. The walls were of insulated pegboard construction similar to the ones in Chicago since that was a tried and true form of treatment.

Later on framed resonator panels were added to various places on the walls after acoustic measurements discovered a tendency for low frequency buildup. Of course, floor transmission was also a concern since we had TV Recorders sitting on the same foundation right on the other side of the wall. To counter that, the sub floor was cut around the periphery of the room inside of the two walls, and Celotex was added to absorb some of the impact vibration and lightweight foam concrete was poured over that to form the floor which was ultimately covered by vinyl tile.

This studio also incorporated a new idea of Bill's to build a "bonnet" as he called it to house the speakers above the control room window. The control room was about 1 foot above the studio floor, giving a good view of the musicians. The doors into the studio, control room and sound lock were also a departure from the usual massive soundproof doors of the time.

Bill had an idea and tried a number of configurations until he came up with the final door design that we used throughout the studio complex. They were made of solid hardwood doors laminated to lightweight hollow-core doors by bonding their edges together with foam machinery rubber strips about 3 inches wide around the periphery of the doors, leaving about a 1/2 inch hollow space inside the sandwich. This worked like the multiple pane control room windows where you use panes of different thickness so that they won't resonate at the same frequencies.

Studio B utilized essentially the same construction treatment. The floor, walls and ceiling were constructed in the same way, but the square footage was only about half that of Studio A. Both studios utilized three Altec Lansing "Voice of the Theater" Speaker systems with modified crossover networks for studio playback. Both studios also used three Altec 604C speakers in proprietary bass reflex cabinets, the standard for all of our studios. All speakers were driven by UA/Dynaco 60-watt tube amplifiers.

**Bob**: My wife and I arrived in Los Angeles in the fall of 1958, and looked up Bill to see about going back to work for him. As Jerry said, he was in the process of building the studio, and didn't have work for me. I worked at various places, including mastering LPs for Contemporary Records, which was a leading jazz record label. I was sitting in for Howard Holzer (who later started HAECO and built mastering systems) who wanted to take an extended vacation. Les Koenig ran the company, and had very high technical standards. I recall his playing an LP that I had mastered, and chiding me for one peak that was 2 dB over zero. I still have the pressings of the LPs that I mastered. Their mastering system was a Western Electric lathe. The cutting system was a Westrex 3A StereoDisk cutter head, driven by amplifiers that Howard Holzer had developed for them.

Roy duNann was their chief engineer. I recall one day we were having problems with the cutting system. We spent some time in fixing the problem, not really sure what we were doing. But we got it running again.

I think it was around April 1959, when Bill called me, and wanted me to come back and work for him. That was the ***third*** time I went to work for him!

Bill had purchased Master Recorders on Fairfax Avenue from Bunny Robyn, and had made him a vice-president of United, and Chief Engineer. I started work there, editing half-inch tapes. I later moved to United, and met Jerry.

As Jerry said, Bill hired Balasz Nagy, later he hired Frank deMedio (yes, the same guy), then he hired Rudy Hill. Frank had just left Cuba with his wife, his son, and the clothes on his back. Rudy was Afro-American. Bill didn't believe in segregation: he hired whomever he felt could do the job, and that went for anyone from porters to engineering staff.

**Jerry:** Frank turned out to be a great choice, since he had been chief of a broadcast company in Cuba, and was familiar with audio equipment and its care and feeding. Frank worked a number of years with us before he left and went to work for Wally Heider on Vine Street and then later started his own console manufacturing business.

## 1959

**Bob**: I remember staying late one Sunday night, as Bill wanted to do ambient noise measurements in Studio B at United. I still recall sitting about 4 feet from the SLM (Sound Level Meter), and reading 30 dB SPL "C" scale on the meter. That was quiet! Of course, if we moved or breathed, it would appear on the meter.

About this time, we started using 3-track machines from Ampex. They were based on a "300" type deck with amplifiers and Sel-Sync head switching mounted above the deck all contained in a steel cabinet on casters. They were hard to move and slightly top heavy. Tricky to move up and down the ramps we had leading into our control rooms.

I remember working a session with Bill where the client (I think it may have been Roger Williams) was doing a pop session featuring piano. Bill was concerned about tape noise, so he talked the client into using a new equalization curve developed by Ampex. They called it "AME," or Ampex Master Equalization. As I remember, the client wasn't very enthusiastic about the idea, as he had heard about overload problems using that curve.

We ended up by doing the session using two three-track machines; one set for the normal 15-ips equalization, and the other for the AME curve. During later playback, we determined that the AME machine was overloaded by the piano. As a result, we did the mixdown from the normal tapes.

On another occasion, Eddie Brackett, one of the mixers, was using a three-track outboard rig to feed the three-track machines, since the consoles were stereo, or two-track. He was having a problem with the rig, and Warren Dace, one of the maintenance engineers was working on the problem. I came into the control room to see if I could be of help. Eddie was rather upset that Warren wasn't working on the rig. He remarked that he knew the rig couldn't be fixed in time for the session, he was thinking about a workaround so that Eddie could start the session.

Rather than running around wasting time in an attempt to fix the rig, Warren was planning ahead. A good example of using his brain, rather than his feet!

**Jerry:** At this time United had 2 fairly large studios, one very small narration studio, a mix down room, and three mastering rooms. Two were mono and one was stereo. All were interconnected by at least 12 tie lines (one row of jacks) through a patch bay located in a central room that later became Studio E, the mix-down room.

**Bob:** It was great for flexibility, but occasionally unsettling if you had a client in the mix down room. Usually, they weren't too upset about the interruption. Since United was the first real new game in town with fresh ideas, it seems our clients accepted our spirit of "try something new." I remember talking with Bones Howe, who is a great mixer. He didn't use much eq, rather relying on a good mix. Later he managed to do an entire session with nothing but RCA 77-DX microphones. That mike was a ribbon unit, with adjustable pickup patterns, all the way from omnidirectional to cardioid. They are much sought after these days.

**Jerry:** That interconnect system gave us tremendous flexibility. By this time our control rooms were all set up for three-channel monitoring, and three-track 1/2 tape.

Bill liked to try out the latest microphones, so we had quite a complement of them. We had microphones from practically all the different microphone manufacturers in many different countries. I remember we were doing a recording session around 5 PM in the evening and the lead singer was using a Neumann U-67 mike. All of a sudden we began to hear a radio station coming over our monitors. The mixer called for a maintenance man to come and check it out and he went through the microphones one at a time until he found the culprit. It was the U-67, which they exchanged for another type of mike and then the maintenance engineer took the U-67 to the shop. But he couldn't make the radio station re-appear.

So he shelved the project until the following day whereupon he tried it again in the same studio and the same input channel. Still no problem. By now he was beginning to think he had made a mistake in singling out the U-67 as the cause of the problem, so he left it in the studio.

The same group came in later that afternoon and started to record. At 5:00 PM sharp there was the radio station. This time it was noted as to which station it was by its call letters. Disconnecting the U-67 solved the problem, but why at exactly 5:00 PM? A call to the chief engineer of the station brought more light on the subject. It seems that the station transmitter was on Catalina Island, and that at 5:00 PM sharp they switched to directional antennas aimed right at Los Angeles. We later found a number of other "capacitor" type microphones that had the same problem. We then had to place small RF chokes in the capacitor's power lead to get rid of the phantom radio station.

**Bob:** Bunny Robyn, (former owner of Master Recorders) now our chief engineer and a legend in his own time for having mastered some of the hottest records around (level-wise and popularity-wise), designed the mono mastering rooms. The equalizer complement was pretty extensive. It comprised a Pultec equalizer, a Cinema Engineering equalizer, a Cinema Engineering HP-LP filter, a Cinema Engineering graphic equalizer, and a limiter. The equalizer, filter and limiter were later replaced with the same items manufactured by Universal Audio.

Both rooms had venerable and reliable Scully lathes, both with Grampian heads. As I recall, they were later changed to Westrex 2B heads.

**Jerry:** The tape machines were Ampex 300's which allowed us to tweak the playback speed slightly to accommodate tapes from other studios who may not have been as meticulous about their tape speeds. Early on, I had modified the tape head covers to enable us to easily adjust the playback head azimuth without removing the cover. It was our practice to put alignment tones at the head of all tapes originating in our facilities so that we could be sure that the mastering or copying machines could be easily adjusted for optimum playback.

**Bob:** The monitor speaker was an Altec 604C in a bass reflex cabinet, which was edgy, but we learned to live with it. If we were given a tape to master, we would spend a few minutes eq'ing the tape. We would note all of the settings on the box, so that if and when the tape was to be re-mastered, the next mastering engineer would know the settings. Cutting 45's and LP's in this fashion gave us lots of experience.

**Jerry:** Bunny Robyn and I designed an EQ log sheet that we had printed and made into pads that were in all of the mastering rooms to enable the engineer to quickly log all of his EQ settings, which were then taped inside the lid of the tape box.

**Bob:** The second floor was the typical wood stage construction. In the back of the building were the tape library, and two mastering rooms, one mono and one stereo. To isolate the lathes from the building, Bill had platforms made. The bottom end was a heavy concrete pad foundation, set below the first floor level. Then four 6 inch round steel columns extended to just above the second floor level. A wood platform was constructed on top of the columns. These columns were braced laterally to prevent any swaying and the whole thing was encompassed within the walls of the room below (also a mastering room). Soft rubber padding was placed around the platform, so styli and other items wouldn't accidentally fall into the crack.

Jerry and I both cut lots of lacquer masters, as did our other mastering engineers, Andy Richardson, Bob Golden, Don Geis, Bill Stoddard and Bill Perkins (yes, the Bill Perkins of jazz fame and a regular with Doc Severinsen's Tonight Show band). The Fairchild and Westrex cutter-heads used an advance ball, made of highly polished sapphire, to rest on the uncut portion of the lacquer. This supported a very small part of the weight of the cutter head, and served to maintain a constant depth of cut. If the ball had run over the lacquer thread, or there was dust on the lacquer, the bit of dust or lacquer would score a line on the lacquer. This wouldn't affect the quality of the finished pressing, but it created a mottled look which was not very pleasing.

Bill Perkins came up with an excellent pun; "What is the difference between state of the art head surgery and an undesirable side effect of using a Westrex head?" Answer: "One is advanced skull boring and the other is advance ball scoring!"

**Jerry:** I remember having a large turbine vacuum system in a little closet that served as an electrical power room next to the first floor mastering room (Studio "D" it was called). It was connected to a little lever switch on each of the monaural lathes (one upstairs and one on the first floor). This provided vacuum to pull the cut thread from the stylus as it was cut from the disk. The vacuum was looped through a jar of water that provided two functions; to gather the thread that was highly flammable, and to wet the thread to cause it to sink to the bottom of the jar. The business end was a short length of ¼ inch chrome tubing that was flattened at the end directly behind the cutting stylus (to fit between the bottom of the cutter head and the lacquer surface, not touching either one). It was mounted to the lathe superstructure and tracked along with the carriage. It was connected to the chip collection jar by a very flexible piece of surgical tubing.

Because this was Los Angeles, the humidity was very low most of the time, and the air moving through the tubing generated considerable static electricity, causing the thread to stick to the inside of the rubber surgical tubing we used to collect the thread from the stylus. When that happened, the thread began to collect on the surface of the disk, and the operator had to do some fancy moves to keep it pushed to the center of the disk with a small artist's brush or blow on it. If it got under the head itself, there could have been a fire on the disk or the thread could have gotten under the advance ball (if that particular head had one).

Neither of these occurrences was particularly desirable as they almost always ended up ruining the disk and sometimes even the stylus. The disks weren't particularly cheap and the styli cost about $15.00 each. I tried many different ways of discharging the static buildup, but because the tubing was rubber and had to be flexible (we were using surgical tubing), it was also a great insulator and resisted any attempts to discharge it. I finally found a company that made specialty tubing for medical applications and they had reasonably flexible tubing made of a carbon impregnated rubber that conducted electricity and we finally had static free thread collection!

Styli didn't last forever either, and at United we all wound our own heater wires on the styli (they were heated to make a smoother and quieter cut) and fastened the little coil of nichrome wire with a dab of dental cement (powder mixed with a few drops of water and applied with a toothpick). Sounds positively tedious doesn't it? Well it was, so it was to be avoided if possible by being extremely careful. But styli eventually got dull causing what we called a gray cut (so called by the look of the groove that was being cut) that was inevitably noisy. Then, if the stylus wasn't physically damaged, it could be recycled and re-sharpened.

The Scully mastering lathes were beautiful pieces of machinery. The lathe bed was made of polished cast aluminum. Within the bed were the solenoids, clutches and gears that drove the main lead screw. It could drive the cutter-head at two fixed speeds and one selectable speed (continuously variable on later models. The turntable was a special cast aluminum platen with several hollow stiffening spokes on the bottom. Those spokes were all interconnected to the hollow lacquer centering pin in the center on top. The top was covered with a thin layer of cork that had thin concentric grooves cut in it. Several small holes drilled through those grooves into the hollow spokes below allowed some of the vacuum to be applied to the bottom of the lacquer thus preventing slippage. The vacuum was drawn through the hollow centering pin through a slip joint on a piece of tubing that shared vacuum with the chip jar.

The turntable was driven by a synchronous motor through a series of belts and pulley/flywheels that could select one of two speeds (33 1/3 or 45 RPM). The speed was selected by a shift lever on the front of the lathe cabinet. Up was 45 RPM and down was 33 1/3 RPM. I made a mistake and threw the lever the wrong way one time while I was mastering a number of new 45's, and accidentally cut one side at 33 1/3 RPM. When the DJ played it he didn't really hear much difference, and since there wasn't anything to compare it with he let it go that way. When the producer happened to hear the tune on the air, he knew immediately that something was wrong and called the record company. Several days had gone by since I had cut the record, so we had to re-cut that side of the record, and while I don't know what it cost us, I'm sure we had to eat whatever the processing costs were to replace the first pressings. This caused Bill to ask me to put lock out switches on the shift levers so that couldn't happen again. I talked him out of that because we sometimes did specialty disks for certain industrial clients that required us to cut 33 1/3 RPM lacquers with 45 size parameters. I promised to be more careful in the future instead.

The stereo mastering room (Studio "G") used a vacuum system made for player pianos, but was really just a standard vacuum cleaner turbine housed in an insulated wood box and designed to be placed in the bottom of a piano which when playing was noisy enough to drown out the motor noise. When placed in the bottom of a mastering lathe, it was too distracting to the operator who was trying to hear the recorded material. We slowed down the motor with a big high wattage resistor (no solid state motor speed controls in those days) and this worked fine for a while. Then the motor overheated due to the decreased airflow through the box and that was the end of that idea. Ultimately we did get it to work through moving the vacuum unit to the attic and running a vacuum cleaner hose and remote switch down to the lathe.

**Bob**: The stereo mastering room utilized a Neumann AM-32 lathe, with a Fairchild 641 cutting system. The cutter wasn't 45/45, which is the physical format for stereo LP's, but a vertical/lateral system. Sum and difference input transformers provided the phase shift to provide the 45/45 format. As a working system, it was excellent, but I recall Frank deMedio and Gil spending many hours tweaking the system. It required large amplifiers running fan cooled transmitter type output tubes. The feedback was provided by a capacitor modulated FM (radio frequency) system. I lost count of the number of lacquers we cut up there, but it was in the thousands (lacquers were cut in multiples and shipped to a number of pressing plants around the country to save on the cost of shipping heavy finished albums across the country).

Bill came up with a technique for 2/3 speed mastering. The tape machine had an auxiliary sleeve fastened over the normal capstan, the tape machine was set to run at 7 1/2 ips. The Neumann lathe could be set to half-speed, so we set the turntable to 45 rpm, then half-speed, which gave us 22 1/2 rpm. Since 2/3 of 33 1/3 rpm is actually 22.222 rpm we were actually spinning about .22 rpm fast and that was taken care of with the capstan sleeve which was calculated to run the tape at 10.22 ips instead of 15 ips. Complicated? Yes, but very effective.

**Jerry:** The setup was time-consuming, but it gave us great-sounding lacquers. Since it's well known that the lacquer medium tends to be pushed out of the way (instead of being cut out) during cutting of very high frequencies due to the higher velocity of the cutting stylus, it is advantageous to slow down the stylus by lowering the frequencies being cut. We couldn't go half-speed, as the bottom end (or low frequencies) was too low for the system frequency response. Since the cutting system is normalized to the RIAA standard and the tape machine to NAB, it was also necessary to pre-equalize the material so that we ended up with the proper frequency balance as heard when playing the tape at its normal speed. The system was primarily used for Reprise Records releases, including Frank Sinatra, Dean Martin, Don Ho and Sammy Davis. I believe we also did one or two of Herman Clebanoff for Mercury records.

Up to this time all our cutting lathes were fixed pitch although they were mechanically capable of running as variable pitch machines (with a few modifications) so we decided that with the length of LP recordings getting longer, we needed to convert at least one of the lathes to variable pitch.

We chose to convert the Scully lathe in Studio "D" (the first floor mastering room) to variable pitch first. This entailed purchasing a kit from Scully containing a servomotor and an amplifier to drive it. These were made to fit right into the Scully lathe console. The pitch selector was belt driven by the servomotor assembly. We had to mount a playback head 15 inches before the regular playback head to provide advance notice to the system of the recorded levels. Under no modulation (tape not running), we would set the depth of cut to 1 mil, and the pitch somewhere around 300 lpi (lines per inch). When the tape was running, the recorded level caused the pitch to expand to a lower number approximately 1/2 revolution before the material was recorded. The louder the music, the lower the pitch (within reason). There were trims to fine-tune the system.

We then did the same for the other Scully on the 2nd floor. When we got to the Neumann lathe with the Fairchild cutter head, it was a little tougher. While the lathe came equipped with amplifiers to drive the pitch and depth functions, we had not been using the system because of the weight of the Fairchild head and the fact that it was resting on an advance ball. In order to use the system it was necessary to increase the strength of the head suspension spring in order to almost lift the advance ball off of the lacquer. The Neumann head suspension unit was designed for a much lighter head, so we had to float the Fairchild almost to the liftoff point and use the advance ball as a maximum depth limit device and suspend the head electro-magnetically to the proper cutting depth using current from the depth servo-amplifier supplied by Neumann. Whew! This made adjusting the depth of cut very tricky (and susceptible to bumps and vibration of the lathe and its platform).

Of course, being stereo, this system required a stereo advance playback head and amplifiers as well as sum and difference coils to convert to vertical and lateral signals required by the lathe. When in action, it deepened the cut as well as changed the pitch depending on the levels in the 90 degree axis of the groove. We later changed the room to a Neumann SX68 cutter head which was a 45/45 system and a much lighter head with helium cooling.

**Bob:** The industry term was 'depth of cut,' but in reality, we were measuring the width of the cut. A misnomer which was regularly ignored by the industry, and probably still is.

I recall one interesting stereo mastering session. This was about the time of the Cold War between USA and Russia. The client had created material to warn people about the dangers of nuclear war, and for the opening, they had obtained a recording of one of the test atomic bomb explosions. I managed to successfully master the material after lots of test cuts. There was lots of low frequency material spread from left to right, with some material being naturally out of phase. That wasn't a simple mastering job! I don't think we charged the client any more, even though the Traffic Office was aware of the singular nature of the job.

Late in 1959, Bill asked me if I and my wife would like to move to Mexico City, as Chief Engineer of a Mexican record company, Discos Mexicanos. My response, even before talking with my wife was, "When do we leave?" For this record company, Universal Audio had built two boards with three-track outputs. Bill had designed the studios for them. He had designed and supplied a mono mastering room, but they still needed a stereo mastering room, as well as a 'gringo' to show them the fine technical points.

### 1960

**Bob:** My wife and I made a trip to Mexico City to see if we liked the city and the company. We did, and moved down there in February 1960. One of my tasks was to build a stereo mastering room. We leased a Scully lathe from Larry Scully, and ordered a Fairchild 641 cutting system. I did the system design, and arranged for the material to be brought down from Los Angeles.

We got word that the Fairchild system was at the airport waiting for customs clearance and the appropriate duty. Alexandro Ghersi, the studio manager, and I went to the airport. My Spanish wasn't too good, so Alex and I agreed he would do all the talking. On the way there, he asked me for clarification as to what the system did. I explained it to him, and he seemed to understand the workings.

We walked into the customs office, and got in touch with the appropriate officials. The system was in large boxes, carefully packed. They had opened the boxes, and we saw the power supply in one box, the cutter amplifier in another box, and the feedback amplifier and the cutter-head in another box. One of the officials took out the cutter-head, which was fastened to a piece of plywood, and proceeded to pass it around for examination. I stood there with my fingers crossed, for if somebody dropped it, we'd be back to square one.

Alex started talking to the officials, gesturing and pointing to the cutter amplifier. I heard phrases from him, which included "fuente de poder" which meant "power supply." I thought about what he said, and realized he was talking about the cutter amplifier as a power supply. After more talking, there was lots of nodding heads in agreement, handshaking all around, and we left. On the way back to the studio, Alex explained that he used the phrase "power supply," which was quite correct when I thought about it. He didn't want to use the word "amplifier" at all, for we would have been socked with 100% duty, as PA amplifiers were being made in Mexico. As it turned out, they classified the system as "scientific apparatus," and only charged us 18% duty.

As I remember, I had one of the engineers do the wiring under my directions. I had great delight in assembling and tuning the Scully lathe, as it came in several sub-assemblies. My experience at Atlantic Records was of real benefit!

Manuel Diaz had been their chief engineer until I got there. At first, there was some resistance about a "gringo" being their chief engineer. I realized later they began addressing me as "Ingeniero," rather than simply "Senor." I guess they determined I knew something about recording.

Four Star Television in Los Angeles was one of the leading television show producers. Herschel Burke

Gilbert was their musical supervisor and orchestrator. He did several music scoring sessions in Studio A, finishing when the record company closed down the studios. Herschel's sidekick was Al Perry. He ran the session in the control room. Since they were doing lots of short music cues, some as short as 5 seconds, stopping for playback would have been very time consuming. After the cue, Herschel would say to Al, "S'all right" with a questioning tone, Al would respond with the same statement, and the session would continue.

I was also consulted when they had some problems with the processing and pressing plant.

One of our daughters was born while we were there. Manuel agreed to be her godfather.

**Jerry:** While Bob was building a stereo mastering room in Mexico, he shipped most of their stereo music output up to me in Hollywood. I would cut the masters and send them back to Discos Mexicanos where they were processed. I got to like Mexican music after a while.

**Bob:** Unfortunately, their estimate of record sales wasn't enough to support their operation, so we came back to Los Angeles about September 1960. Bill didn't need me at that time, so I worked for another studio, Universal Recorders (no relation to the Chicago studio) as maintenance engineer.

I did regular routine maintenance on the tape machines, Ampex 350s, 351s, and 300 machines. The studio was a subcontractor for production of LP masters for AFRTS (American Forces Radio &

Television Services. The shows were 30 minutes on a 12" LP.

The production manager played a test pressing for me that slowed down in spots. Since the tape had no problems, I clarified for him that the problem lay either with the Scully lathe speeding up or the tape machine slowing down. I determined that it was virtually impossible for the lathe to speed up. The problem hadn't been discovered until about three weeks after the lacquer had been cut, so I had no way to determine which machine was at fault.

It turned out that once a mastering session was started, the mastering engineer would go to the production office to sort out the next tapes to master. After the problem, the production manager issued a memo that mastering engineers had to remain in the mastering studio while the LP was being cut.

While at that studio, I started to build a box that Gil had designed at United Recording. It simplified for everyone the process of the daily check of the mastering systems. This check would verify the system frequency response. The box was an equalizer with the RIAA playback curve. When you looked at the test disk, you would see a flat response curve instead of the normal preemphasis curve. The flat response was much easier to view.

The studio was a subsidiary of another recording studio in Los Angeles. One day, their manager and chief engineer appeared on one of their infrequent trips. They asked what I was building. I explained the idea to them. Not being familiar with the idea, they told me to stop the process.

In early 1962, I think it was, Bill called me to come back to work for him - for the ***fourth*** time! He had bought another studio, which Jerry will begin talking about. Bill again wanted me to coordinate design and construction, this time for a complete studio renovation.

Here are pictures taken by Jerry and Bob through the years. The dates are when the pictures were taken.

1954: Ampex 3200 Duplicating system installed at 111 E. Ontario.

1954: Dale Harrington in Universal's dubbing room at 111 E. Ontario, Chicago. Note the Ampex 300 in front of Dale, as well as the custom-built dubbing lathes behind him.

1954: Dale Harrington checking levels on VU meter.

1955. Bill at Studio A console at 111 E. Ontario. The machine is an Ampex 350. They didn't arrive until 1955. The vocal booth at right was on casters, and could be moved (with work).

1954. Bill at Studio A door. The control room window is at the right.

1955. Bill in meeting in Studio B, talking with Mal Chisholm, Tom Parrish, and Bob about the design and layout of Walton Place.

1955. Mal Chisholm in meeting. He later taught at Columbia College in Chicago.

1955. Tom Parrish in meeting.

1955. Walton Place construction.

1955. Walton Place, looking east, early construction phase.

1955. Walton Place, looking down to Studio A.

1956. Universal Studio A (Chicago, Walton Place). View of SE corner from elevated control room

1956. Universal Studio A (Chicago, Walton Place). View of NE corner. Note moveable isolation booth at top left.

1956. Universal Dubbing Room (Chicago, Walton Place). The five-headed monster.

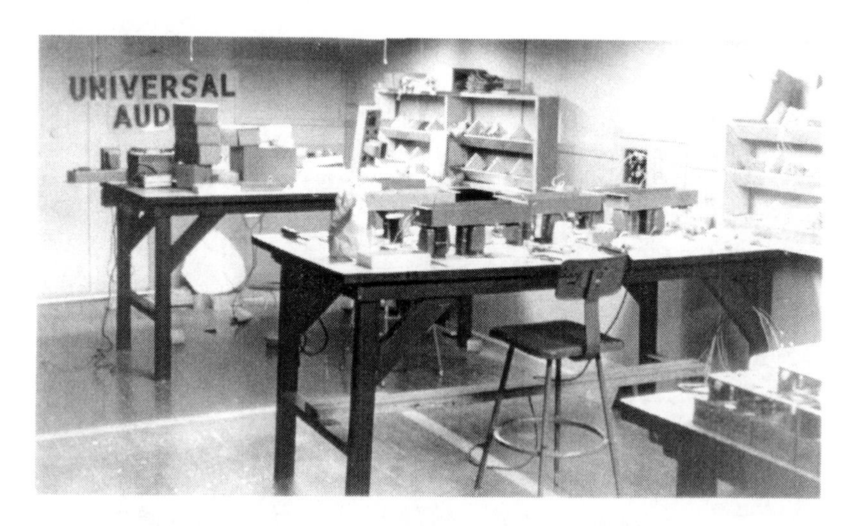


1958. Universal Audio in the early days showing the 2nd floor "production line". On the bench are UA/Dynaco power amplifiers being assembled.

1958. Long view shows UA 100 D preamplifiers being assembled (on bench in foreground) for the new consoles.

1958. Universal Audio in the early days showing the 2nd floor "production line". On the bench are patchbays being wired.

1958. United Studio A (Hollywood). View of control room during construction shows "bonnet" for control room speakers.

1958. United Studio A (Hollywood). View along West wall shows saw-tooth layout to prevent direct reflections from opposite wall.

1958. United Studio A (Hollywood), another view of "bonnet."

1958. United Studio A (Hollywood), Acoustical panel frames.

1958. United Studio A (Hollywood), View to SW corner.

1958. United Studio B (Hollywood) showing control room under construction. Note the speaker "bonnet" above control room window.

1958. United Studio B (Hollywood). West wall showing saw-tooth layout. The inset panels are low frequency Helmholtz resonators to control room acoustics. The rectangular opening is an air-conditioning duct.

1958. United Studio B (Hollywood), Control Room wall, bottom of "bonnet" overhead, before trim and glass installed. Note the separate walls below the "window."

1958. United Studio B (Hollywood), View to SE corner. Not all walls were saw-toothed. Belden shielded wire in foreground, ready for pulling.

1959. Jerry Ferree and Balasz Nagy, United Studio A.

1962. Bud and Ethel Morris. Bud is showing off his Xmas present.

1962. Gil with RC model airplane, Rudy Hill next to him.

1962. Mildred and Tony Parri, who is showing off his new golf cart.

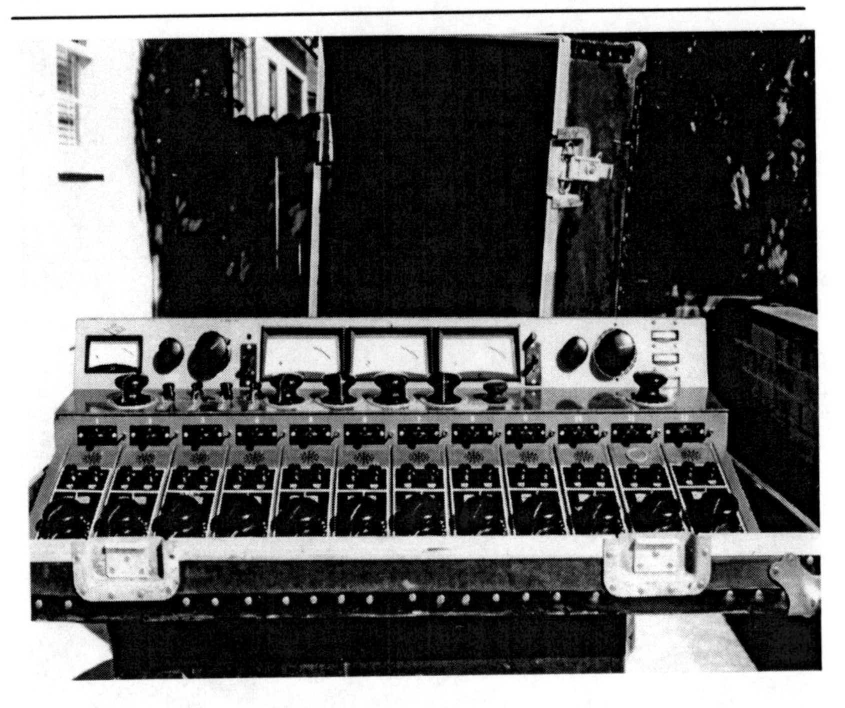

1963. 310 Remote (portable?) console.

1963. 45 degree view of remote console.

1963. Inside view of remote console

1965. Hal Halvorsen on fishing boat, rebuilt by Hal and Bob Doherty.

Chief Engineer Jerry Ferree and New Closed Circuit TV Film Unit

1965. Jerry Ferree with telecine/transfer machine.

1966. Bud talking with Bill before the DISC meeting.

1966. Bob Doherty talking to the group about United and Western.

1966. Bill talking to DISC investors.

1966. Bud Morris talking to the DISC group.

1966. Bob Doherty talking to the DISC Group.

1966. Jerry Ferree, talking to the DISC Group

1966. DISC meeting, not a large group, but meaningful.

**1962**

**Jerry:** In 1961, Bill had bought Western Recorders (1 block east of us), and annexed their staff to ours. We then started to remodel the Western building that had originally been a broadcast studio. Bud Morris (who succeeded Bunny Robyn as Chief Engineer) was put in charge of the renovations. They had one big auditorium with a stage, and two smaller rooms. We started by remodeling the smallest room which was about 2/3 the size of Studio B (the midsize studio at United). This was named Studio 3, and was used for rock & roll recording. The name later became famous when Wally Heider (who, although a well known amateur jazz recordist, got his professional start at United) built a studio on Cahuenga Blvd. copied after Western Studio 3 and built by the same builder. He named it "STUDIO 3".

**Bob:** After Studio 3, Bill went to work on the auditorium. It was named Studio 1. There were about 400 seats in the auditorium. I recall putting an ad in the LA Times; we sold most of the seats to a small church that was delighted to get them. After removing the theater seats, and removing the stage floor, we had a huge area to work with. Hal Halverson was the carpenter Bill hired for the remodeling. After the seats were removed, Hal had the sloping area leveled with decomposed granite. I think we ran conduits on top of the material, and then a concrete floor was laid.

**Jerry:** Part of Bill's acoustic treatment ideas for Western Studio 1 was to create areas of controllable liveness. To this end he created large moveable shutters that could move up or down exposing a soft side or a hard rounded surface. There were three of these shutters on each panel around the room (a total of 14) each 2 feet high by 3 feet wide suspended horizontally by piano hinges and tied together by 1/8 by 3/4 inch aluminum strips on pivots at each unhinged corner. Needless to say these were very heavy.

Bill and Bud went to a surplus store one weekend and came back with 48-volt aircraft winches and 3/16-inch stainless steel rope. He had the carpenters build a box under each panel and place a pulley at the top. Then he turned it over to me. "Make it work," he says. I tried. I made up a panel with a three-way reversing switch for the motors and a BIG power supply using a transformer he found, and discovered that these things drew 20 Amps at 48 volts! Moreover, they didn't stop on a dime and they coasted after you switched them off. That put slack in the line when going down that sometimes caused the wire to come out of the pulleys, and going up LOOK OUT! The line snapped and the whole thing came down with a resounding crash.

I tried putting in limit switches for both directions, but while that worked most of the time, it wasn't foolproof. You didn't want to be standing less than two feet away while it was moving. The big problem was that they were just too heavy for this kind of thing, and the panels moved too fast near the end of the cycle. If you put resistors in to slow them down, they never made it past the middle of the cycle where the weight was greatest. We finally abandoned the whole thing and fastened the panels permanently in place after trial and error testing of the sound around the room. Hal Halvorsen was a great rough carpenter, but needed some help for the finish work. He hired two carpenters to work with him, between the three of them, the work really moved along.

The last studio to be renovated was Studio 2 and was relatively uneventful. It contained an enclosable alcove with fairly dead acoustics next to the control room that was used as a vocal booth. I tried out an experiment using self amplified headsets connected to a pickup coil which sensed the field of a couple of turns of wire wrapped around the studio about ten feet off the floor. Like a big transformer primary driven by a high-powered amplifier. It was a crazy idea that never worked well enough to take hold. In those days we didn't have high-energy infrared diodes that could be modulated, as that would have been an ideal use for them. Several of the percussionists were using a low power RF system put together by Larry Bunker, a notable percussionist of the time who was doing a lot of film work. Those guys were severely hampered by being tied down to a headphone cord, as they covered a lot of ground behind their percussion instruments.

**JERRY:** About March, Bill was again looking for more areas to expand into. United purchased Coast Recorders of San Francisco, which later was renamed United Recording Corporation of San Francisco. It was located in an old live theater building on the North side of town on Bush Street, and had never been remodeled after being turned into a recording studio. After operating there under the management of Don Geis (one of United's original mastering engineers) for about a year, Bill began looking around for another location in San Francisco.

**Bob:** About this time, Bill hired Frank deMedio, initially to pull audio wire. He turned out to be a great help, and Bill later asked him to work with Warren Gilman in studio maintenance. When Gil moved to

Universal Audio, Frank took his place as Chief Maintenance Engineer. He later left United and went into business building custom consoles. He's still very active in the professional audio field in Los Angeles.

I recall Jerry and I pulling mike cable for the studio. In those days, we used Belden 8761, 2-conductor #22, shielded, with wire numbers, of course. As always, Bill gave us suggestions and guidance as to the best mike panel locations.

**Jerry:** Bill liked to try out all the trendy mics of the time. We had to make up adapter cables to go from XL3 to the Hubbell plugs. This was because most of the mikes of the day came pre-wired with XL's. I remember later uninstalling those Hubbell microphone connectors as Frank and I and Bob Felthousen changed the whole plant over to Cannon XL3 connectors. It took a week or more, and that meant having some of our mikes with one connector and some with the other connector while the studios were changed one by one.

**Bob:** Bill was great for crash programs. He had been in the Army, and I don't know how he lasted, because he didn't stand around and wait. Jerry and I recall a couple of all-night stints where we later crashed in a nearby motel, courtesy of the studio. I think we saw them coming, but we weren't about to say no to Bill. All in all, he was a fascinating guy to work for.

**Jerry:** All of these studios with the exception of Studio 3 had 18-foot ceilings with diffuser clouds. Studio 3 was in an added on area of the building and only had a 12-foot ceiling. The playback speakers in all of the studios of both buildings were modified Altec A7 systems. The control rooms generally had Altec 604D speakers in bass reflex cabinets.

Bill favored having the speakers in front of him in the control room so he always designed the control room with a "bonnet" above the control room window that jutted out into the studio. It was into this that the monitor speakers were mounted, and held in place by safety chains. In Studios A & B, the console was set up against the window wall. For Western, I suggested that we raise the console about a foot above the studio floor and set it back just far enough for a row of chairs (we used some of the leftover theater seats from the Western auditorium) in front of the console at ground level. This allowed the mixer a better shot at the speakers, and visitors a better view of the studio. Bill liked the idea, and we built the three control rooms at Western that way.

Western also had a small narration room (we named it Studio 4) that was used for recording spots and voice over. There was also a mono mastering room originally, but it was later removed and became the control room for a mix-down room. As I recall, the mastering system was moved to our Las Vegas studio to cut dubs.

United and Western were tied together by dry telephone trunks provided by the telephone company. There were 24 pairs as I recall. About 16 pairs were dedicated to feeding to and from United and Western. They required equalized amplifiers at the send end. Others were dedicated to feed off-premises echo chambers located elsewhere in Hollywood. There were a number of people who had built echo chambers for hire in those days. Some were pretty good and others were either tinny sounding or heavy with extraneous ambient noises.

**Bob:** As part of the Western renovation, I became involved with the air conditioning system. I think that Bill said, "Take care of it." Unlike the United building, which had separate units for each studio, the Western building had two units. We needed a contractor to handle maintenance on the system. I put together bids from ACCO in Glendale and another contractor for Bill, Tony, and Bob Doherty's approval. ACCO turned out to be an excellent choice.

In the Western building, there was one central unit that handled the studios on the first floor and rented offices on the second floor. The system had been well designed, provided high-volume, low-velocity air to the studios. However, the tenants on the second floor were continually adjusting the upstairs thermostat. After talking with ACCO, I had them install a thermostat in the return air ducting for the second floor, but leaving the existing thermostat in place. Now the tenants could adjust the hall thermostat as they desired, but the temperature stayed constant. We never got any complaints after that.

In Studio A and B at United, the musicians were constantly adjusting the thermostats. As a result, the main cooling coils would ice up, requiring the system to be shut down. I don't remember if the idea was Jerry's or mine, but at any rate, I put locking covers over the thermostats, and screws to hold the adjustment levers in place. After that, we didn't have any more problems.

Bill's style of management was not micro-management, far from it. He gave you the project, checked to see the progress, and that was that. As Jerry mentioned, he was not strong on praise, but you knew when you had done something well.

**1963**

**Bob:** I recall getting to know the guys in the local Pac Tel office quite well. They knew we were serious about good quality audio and that kept us in good stead with them. I recall some of the engineers complaining that the tie lines to Western were noisy. I called the Pac Tel office, and they sent over one of their people with a fancy transmission line tester. Mind you, this was 1963, and the term "digital" didn't exist. They tested the lines, and found they were all about 0 dbrn, which was as good as it gets. Their reference was "db above reference noise," which was -90 dbm in our world, very quiet! After further investigation, I found that the engineers weren't driving the lines at an adequate level and that the residual noise from the associated electronics and the ambient noise in the chambers were the probable cause. I advised the other engineers as to Pac Tel's comments, and suggested changes.

Somewhere about this time, in order to increase the capital worth of United, Bill and Tony created the "Diversified Investment Study Club" or DISC. I remember that Manny Berman, who was Bill's accounting firm from Chicago, was there. Individual people, employees and working musicians, contributed funds to DISC. The fund was invested in United Recording. My wife and I invested, and some of the studio musicians invested quite a bit. I don't remember if we made any money, but it was a fun group, nevertheless.

About a year later, the Smothers Brothers wanted to edit material for a new release. It was called "Mom Always Liked You Best." Nook Schreir and Tommy Smothers had worked with Jerry, but he was on vacation, and they wanted to start immediately. Dick Sexty, our Traffic Manager, asked me if I wanted to work with them. Nook remembered me from Chicago. For the next week, from about 10 AM, with stops for lunch and dinner, then working until I hollered 'uncle,' I edited the 3-track, half-inch tapes of their various shows in Las Vegas and Los Angeles. Nook wanted to track the various edits, which sometimes got very involved. Tommy Smothers was a great and funny guy to work with.

Jerry came back from vacation just before we were doing a studio session to add four cuts to the LP. I hadn't done that much mixing; fortunately Jerry showed up at the right time, and got the mix started for me.

**Jerry:** Actually The Smothers Brothers almost always recorded on 3-track tape. We put the boys on the outside tracks and put the bass and guitar in the center. We usually miked the audience with at least 4 or 6 mics, and recorded them in the mix on the left and right channels. Almost all of their album material was recorded on remote. I ran two 3-track Ampex machines, and we had an eight channel modular mixer with basic EQ reverb sends and receives and three output channels. We had a rack containing a couple of UA limiters and a couple of UA power amplifiers and the power supplies for the console.

We tried to mike the audience from above, but if the cable runs were too long, (i.e. ceiling too high) we would use lighter weight and more sensitive condenser mics. Most of their albums were compilations from several performances. We recorded in college gyms, concert halls, coffee houses, pizza clubs and Las Vegas hotel shows. Nook Schrier was always there to preside over the recordings. I'll never forget I rolled the company van in the middle of the California redwood forest on one of those trips. It was late at night and pitch black. I took a turn a bit wide and the right rear tire dropped into the soft shoulder. I got into a long slow skid when it popped out and we went over slowly into the brush on the opposite side of the road. Fortunately neither of us was injured, and none of the equipment was damaged significantly, so we had the van righted by a tow service and drove on to the next venue. Boy, were they shocked when they saw the condition of the van.

**Bob:** Bill had a shot at doing some magnetic film work for a client. He asked me to get a 35 mm mag film recorder. Magna-Tech would build one in about four weeks, but Stancil-Hoffman would build one in about a week. My thoughts were to wait for the longer delivery machine, which I thought would be a better machine. Bill wanted the machine immediately, so I placed the order with Stancil-Hoffman, who was located nearby in Hollywood. As it turned out, they were a week or so late, and Frank deMedio and I spent some time getting the machine somewhat near specs. But only somewhat! I'm not sure the client ever came through, but then I don't recall Bill chewing me out for late start-up of the machine.

**Jerry:** Well, the Stancil-Hoffman machine was, as Bob has said, a disaster waiting to happen. The wow and flutter was horrible and Bob and Frank worked a long time trying to get it working acceptably. Since time was running out, I borrowed a 3-channel Westrex film recorder (mind you we are talking 35mm here) from our friend Larry Aicholtz of TV Recorders next door. It was mounted in a "portable" case and placed on a roll-around table, so we could roll it into the control room. It was HEAVY! It soon became apparent that it was also very noisy mechanically, and the decision was made to put it in another room with the recordist and use an intercom.

The client happened to be Mercury Records, trying to jump on the 35mm bandwagon started by Bob Fine at his film studio in New York. The artist was Herman Clebanoff and his string heavy orchestra doing Broadway hit tunes. You can imagine what that would have sounded like on the "wow machine" Frank nicknamed the "Standstill"-Hoffman. The idea was to master directly from the 35mm film, but the 35mm machine was too heavy to carry up to the second floor stereo mastering room, so we did a "live" mix-down from the mix-down room and patched the audio to the stereo mastering room upstairs. The mastering engineer would start the master lacquer, and tell the mix-down engineers on the phone who then started the Westrex and proceeded to mix the album non-stop. Whew, talk about the mother of invention! After the second or third album like that, I had Hal (our carpenter) install an eyebolt in the stairway ceiling so we could hook on a block-and-tackle rig to hoist machinery up to the second floor. We used that many times!

Ultimately we did get the Stancil-Hoffman working properly, and I installed a sync-slave motor in it also, so we could lock it to a projector. I'm not sure how it came about, but we started to get some film scoring business and some of the producers said they liked the sound we were getting and if we could project film, they could bring us a lot more business. Our major client was Four Star Pictures who produced the Outer Limits TV show, usually scored and conducted by Dominic Frontieri. I talked Tony Parri, our comptroller, into letting me buy a used 35 mm theater projector. I found an old Motiograph projector that I rebuilt with a sync-slave motor to drive it and we mounted it on the second floor between the two studios at United. Bud Morris and Bill had a habit of hitting all the second hand equipment shops that were around Hollywood at that time, and scrounged up a used "distributor" set somewhere.

We mounted it in a small room that housed the electrical panel for our half of the building, and I took on the task of designing and building a control panel for it. A "distributor" is the film industry's name for a motor generator set that produces the synchronizing source to run projectors and film recorders in exact synchronism. Because of the huge inertia involved in getting the film machines up to speed, it had to be done "softly" or you tore out all the sprocket holes in the film. The whole system is called an "interlock."

We used this setup to record a number of scores for films and television shows, but there was a lot of down time and lost effort setting up the studio so that the musicians could see the screen, setting up lights on all of the music stands (you can imagine the web of light cords all over) and then setting up the microphones around all that. In addition, most of the time the musicians had to have headsets so they could hear a tempo track that was being played back on a synchronized film playback unit (dubber). This added more cables to the mess. I was not happy, the setup guys were not happy, and Dick Sexty had to allow much more time to set up the studio so he was not happy. It was apparent that we were just not set up to do this type of thing as the film studios did it. Their scoring stages were set up for one function, whereas we had to change from scoring to recording. It was obvious to all that we had to have a better system. It was a long time in coming.

**Bob:** This isn't directly about Bill, but indirectly because of his planning. Dick Sexty, the Traffic Manager, advised us that we had a potential client who wished to build some pseudo three-track stereo material. As I remember, the client was a film production company who had some unusual choral material. They had been quoted high rates by one of the film post-production houses, and wanted to see if we could lower the cost.

Jerry and I put our heads together in the three-track mix down room, spent some time in experimentation, and then advised Dick we could do the project. He came up with an hourly rate. As I recall, we did a fair amount of work for the client.

**Jerry:** As I recall, we recorded the material on film and played back three versions in sync, but slipped by a few sprocket holes each. Add a little reverb, et viola! You have a stereo choir.

**Bob:** I remember that several clients wanted pseudo-stereo material made from monaural. We became rather adept at the process, using the mix-down room at United. Since we had a great collection of equalizers, filters, and limiters, the only limitation was our ability to create an acceptable pseudo-stereo sound. We finally came up with a system using a high-speed tape loop (to create a short time delay) and a combination of equalizers to separate high and low frequencies. This produced a passable pseudo stereo effect and we used it on several recordings.

**Jerry:** In mid 1964 I and Bud Morris (at that time United's chief engineer) and I surveyed a number of buildings in San Francisco that were available for lease. We would go in and measure the ambient noise floor in various parts of the building for traffic noise, airplane traffic, helicopters, etc., make physical measurements and so on. We looked at a number of buildings large enough for studio facilities, many of them old warehouses.

United finally bought some property south of Market Street and shared the building with Francis Ford Coppola's Zoetrope Studios. There, under Bud's direction, they built one fairly large studio, and one smaller one that became popular with the rock bands. There was also a manual Scully lathe brought from the Bush Street location for the mastering room. Until the move, most of the recording in the San Francisco Studios was advertising.

In July 1963, Bill built another studio, this time in Las Vegas. It was named United Recording Corporation of Nevada, or URCON for short. It was equipped with the old console taken from Western Studio 3 when it was upgraded. The Los Vegas studio was managed for a while by Jack Eglash who just happened to be the business manager for the Los Vegas musicians union. He also had a show band at the Dunes Hotel. A kindly gentleman by the name of Walt Payne did mixing and editing chores. Bill would fly over whenever any of his "artist" buddies wanted to record in Vegas. Walt later moved to the San Francisco studio.

**Bob:** What Jerry didn't mention was that the old board from Studio 3 was built around the UA 310 input module. Universal Audio (now run by Bill's son) has resurrected a later version of them, the UA 610, and is now manufacturing them.

Some people that I met really made an impression. I was walking out of the front lobby at United, and passed a gentleman sitting and waiting for someone. I think he had just finished a session. I said "Hello" in passing, and very courteously he said "Hello." The gentleman was Fred Astaire.

**Jerry:** One interesting anecdote about the studio in Las Vegas; it was built in a complex on Industrial Road directly behind the Stardust Hotel. Right behind the building (about 50 feet) was a train track. The train had a bad habit of going by right about the time you were in the middle of a recording; and even if you couldn't actually hear it, the ground shook so hard it rattled the mike booms and music stands. Bill arranged with the musicians union for a special dispensation for trains called "train time" which could be deducted from the session time. Then, still not satisfied, Bill ran a wire out to the train rail and connected a sensor that would ring a buzzer in the control room announcing the arrival of a train.

While Bill had done this all under the cover of darkness, being sure to cover the wire with a couple inches of sand (it's desert after all), he didn't know that the thing was throwing off block signals for miles in each direction. Pretty soon the rail guys were out looking for the ground fault that was screwing up the signals. They went back and forth over the area in their little test car full of electronics, but they couldn't see the hidden wire. They pinpointed the spot where it was attached, and jumping down from their little car, they found the wire, pulled it up, and guess what? It led right to the back door of the studio. They weren't too happy to find out that Bill put it there, but when he explained why he had done it; they let him off with a warning, explaining that tampering with rail signals was a Federal offense, and ripped out his wire. Boy, was he embarrassed by that one! After that, they just relied on "Train Time" for compensation.

## 1964

**Bob:** Bud Morris, who had mixed his share of sessions through the years, and had been recently mixing Lawrence Welk sessions, had moved to Universal Audio as general manager. The previous manager had little concept of finances, and Bill asked Bud to take his place. Bud was doing the sales work for custom boards, which were attracting some interest. I started working with Bud on sales, then started to learn about the custom console design business. Warren Gilman, whose nickname was "Gil," was a great teacher.

At that time, UA was in the Western building. In addition to manufacturing the 1008 mike preamp and the 175 limiter, we were building custom boards. We had designed and built a large sound reinforcement board for the Desert Inn in Las Vegas where Ed White was the sound director. Bud Morris and I were working closely together in getting console projects. We did a small recording board for Chuck Blore Creative Services that I installed for them, with Bud's permission.

**Jerry:** During 1964 we phased in a couple of Ampex 1/2" four track recorders. They were pretty much the same as the three track version, but with another amplifier and an additional Sel–Sync switch. Up to the time we got the three tracks, most of our recorders were mounted in our own proprietary cabinets on casters for easy mobility. The three and four tracks were in Ampex upright steel cabinets and not easily moved to the various studios at United let alone up the street to Western as was sometimes necessary. We designed a special hand cart with large pneumatic tires to move machines between buildings (approximately 1/2 block and up and down two curbs). Needless to say, we avoided moving them whenever possible.

**Bob:** Around this time, Bill was doing lots of sessions for Reprise Records, which was a new label started by Frank Sinatra. His sessions were an event in themselves. Usually we had studio guards, who would control entry and allow in those people Sinatra had invited to his session. Normally, in almost any session, there were no visitors, as this was simply work. But Sinatra liked singing to a group. Quite often, there were 20 of his friends invited to his sessions. Bill mixed the earlier sessions, and then Lee Herschberg engineered the later sessions.

Frank Sinatra asked Bill if he had any people who could work on music systems (called Hi-Fi systems in those days). These systems were installed in homes owned by Frank's friends. Bill asked me if I wanted to do the work. That meant contacting the owner, determining what should be done, and supervising the work. I said 'of course,' so he gave me the first job.

I contacted the owner's secretary, and arranged to be at the owner's house. The house probably was in Bel Air. When I drove up to the area, it was obvious that these were mansions, not houses. The secretary (a neat guy) met me at the door, then let me admire the entrance hall, which was large! I saw various paintings and sculptures, then he led me into the owner's library. Above a fireplace was a Van Gogh painting, one of his "Sunflowers." I stopped to look at and admire the painting, then mentioned to the secretary that I didn't think the painting was a copy. He said it wasn't.

The owner was William Goetz, who was a well-known figure in the film industry, and his art collection was also well known. The system in the library had been installed some years before, with monaural sound. I don't remember how we finished the system, but apparently Mr. Goetz and Frank Sinatra were happy.

A month or so later, Bill gave me another owner to contact. This individual was Gene Kelly, who lived in Beverly Hills. I met him briefly, a very pleasant person. Again, the system was monaural, and again I don't remember the details.

Bill gave me another yet another owner to contact. I arrived at their house in the late afternoon, and looked at the owner's system. She asked me if I wanted a drink, I said yes, so Natalie Wood mixed me a drink.

I was very involved with console design at Universal Audio, so I put off working on Natalie Wood's system. About a month later, I got a call from Frank Sinatra, chewing me out for not working on her music system. I assured him that I'd immediately get to work on the system. As I recall, Frank deMedio finished the work. As Jerry has met the same level of personalities, we both knew these people were working, just like us, but they were paid a bit more!

**Jerry:** Sometime in 1964 (I'm not sure when) 3M Company began to show their new 2 Channel Tape Machine. It was an entirely new approach to tape handling, and we were very impressed. Once you knew how it worked, it was extremely easy to set up. In addition, the electronics were on computer type cards and could be easily exchanged for service and were mounted in the cabinet below the transport. The level meters were on a small panel above the rear of the transport. All controls were fixed (screwdriver adjusted) and easy to get to.

We didn't order any as yet as we had a full complement of 2-track machines already.

**1965:**

**Bob:** Bill had asked Bob Doherty to join United and Western. I had known Bob in New York, where he had been Chief Engineer of MGM Records. Bob often asked me to be second engineer on his sessions, as we worked well together. I'd do it as a favor. One time at United in Studio A, Bob was doing a session for a major label. The artist was Jack Jones, who was doing "People" (I think it was) along with three other songs. A well-known individual was the arranger and conductor.

Unfortunately, the conductor was involved in a contract dispute with the label, so he wasn't too happy. His conducting was very good, but the producer wasn't being helpful. If he thought something wasn't right, instead of making suggestions over the talkback, he'd simply say "Take 4, let's start again." Since both the conductor and Jack were professionals, this didn't go over so well. As the song was a rubato tempo, the conductor should have been conducting the orchestra around Jack. But he wasn't very happy, and he wasn't working with Jack, just conducting at a fairly fixed tempo. As the takes progressed, Bob and I looked at each other, waiting for something to explode.

Finally, it did! Jack was towards the back of the studio. Suddenly we heard a crash, and Jack hollered "I can't deal with this!" Suffice to say, the conductor stopped the orchestra, and everybody looked at nothing at all. I waited, and then from a nod from Bob, I stopped the tape machines. He suggested I set up Jack's music stand that he had knocked over. So I went out into the studio to set up the stand. It was one of our special stands made of nice hardwood for the better artists. Unfortunately, it had broken, so I rounded up another stand and went back to the control room. By this time, tempers had gone off the boil. The producer was more amenable, the conductor realized his immediate obligation, and the session continued much better than before.

At Western, we had one room that we were going to remodel into a dub room. The recording lathes had been in place before Bill bought the studio. I don't remember the manufacturer, but they were just barely adequate. Since the lathes were around the outside of the room, it seemed a good idea to put the tape machines in the center of the room. This necessitated a column and cabinet where the machines could be patched in. Also power was available there. As I remember, the cabinet in the center was dubbed "The Periscope," and was the object of some derision.

**Jerry:** I remember those lathes. They were bare-bones copies of Scully lathes by some small time machinist. I think they were called Arcturus. Pretty good, but not great. There were originally six of them, but they were in a sad state of disrepair, mostly bearing problems. They were originally used to record radio transcriptions and only ran at 33 1/3 RPM. One of the Western Recorders engineers had modified one of them to run at 45 RPM also. I think it was Don Blake, one of the original owners. When I was asked to check them out, I found bad spoke patterns (rumble) on most of them, and we finally got parts from several of them to make two good ones. I believe they had Olsen cutting heads on them. The rest of the pieces went into the back room storage pile.

We eventually installed a Lyrec 3-speed motor on one of the lathes (the type used on the Neumann lathes).

We also bought a couple of player piano electric vacuum units to pick up the lacquer chip, and after burning up a couple of motors (due to insufficient air flow), we put in resistors to cut down the current (and the noise). These lathes weren't used all that much as there was a perfectly good Scully lathe right down the hall in the mastering room. Also, it took a certain finesse to operate those lathes. Since there was no automatic run out or head lift, so you had to do all of that manually. One slip and there went the stylus!

**Bob:** Because more room was needed for the Traffic Department, and there was no more room available in the United building, it was moved to the Western building. Bob Doherty, who was appointed Studio Manager, came up with a neat idea for scheduling. By this time, we had six studios of various sizes, four mastering rooms, and one mix-down room. Live sessions were still booked in three-hour segments, as dictated by the musicians union. Studio setup and breakdown had to be scheduled, as the layout seldom was the same from one session to another. Bob designed a scheduling sheet that encompassed all rooms at United and Western for one day. The sheet was about 15 inches by 25 inches, not small!

These were laid out on a long fixed easel, about 12 feet wide, the width of the main room. There were five sheets immediately available, so the entire week could be seen. We weren't closed on weekends, but the activity was considerably less. The sheets were moved at the end of the day, so the current day was to the left. Bob Doherty also had several telephones placed in strategic locations, so it was fairly easy to check a sheet. A group was set up to run the office; Dick Sexty was Sales/Traffic Manager, and there were three ladies in the Traffic Department; Leila Greenstone, Betty Belson, and Lena diFiori. They were responsible for scheduling the clients and assigning engineers. (We provided them in those days).

As I was working with Universal Audio and Warren Gilman most of the time, Bob Doherty and I would trade comments about studio work vs. console manufacturing work. At that time, United and Western provided much more revenue than Universal Audio. Later on in 1977, when Bill sold the studio operations to Allen Sides and UREI to Harman International, I imagine that UREI was the lion's share of the sale.

**1965 (still)**

**Jerry:** When Bud left his post as Chief Engineer to move to UA, it was taken over for a short time by Al Butow (previously from another studio), but things didn't work out as well as Bill had hoped, so he left and I got the job. Tony Parri probably prompted that decision.

I had been doing quite a bit of cartoon voice recording and editing for our neighbors next door, TV Recorders. They did a lot of dubbing business for Hanna Barbera and other small studios in Hollywood. Recommended by them, I did quite a number of score mixing sessions for TV cartoons. I asked Bill if it was all right to use the small narration room after work to edit cartoons for TV Recorders, and whether he thought it would be a conflict of interest. He said he'd get back to me, and later said he thought it would be all right since it was also bringing in work for our studios. I got to be on a first name basis with Bill Hanna and Joe Barbera, as they often directed the voice recording we were doing. One day Bill Hanna said he had been talking to Larry Aicholtz, the owner of TV Recorders, and he had told Bill about my involvement in the building of United's studios. He told me they were making plans to build a new facility in the Cahuenga Pass, and would I be interested in designing a dubbing studio for them. I was totally taken by surprise, but yes I was interested. I told him I'd have to think it over and let him know. Again I checked with Bill Putnam to see what he thought and that I could get us another console to build while building my own reputation as a designer. I would do all the work after hours in the evenings and on weekends. Bill said he thought it would be good experience for me, and as long as it didn't interfere with my work at United it was O.K. with him.

When I told Bill Hanna I would like to do the job for him, he immediately booked an appointment with himself and his controller to meet with me over breakfast at one of his favorite restaurants. We met on a Saturday morning and went over all of the things they wanted to include and be able to do in their facility. We met again later and I gave them my proposal to look over. What nice people they were to work with! They made a few changes to the proposal and we were off and running. I hired Bob to help me and we proceeded to design a dubbing/live mixing console based on our ideas about the business. I showed it to Larry Aicholtz as I valued his opinion as a seasoned film engineer. He made one or two suggestions which I incorporated and we submitted the plans to Universal Audio for a bid. The Console was to use UA's pre-amps and program amps, power supplies and speaker amplifiers. It had to be able to record live music and narration as well as act as a two-man dubbing console. It had to be able to control the recording equipment and be run by one man if necessary. Meanwhile, I recorded the score for H & B's TV special *Alice in Wonderland*, and later full length features *Hey There Yogi Bear* and *Our Man Flintstone.* Bill Hanna's controller set up charge accounts for me at several electronic and film supply houses in LA and I had ordered the 35 mm film equipment from a new company in New York who were eager to work with us. Their name was Magna-Tech, and they built very good equipment for us. I later ordered some for our Western Recorders film room also.

Bill Hanna introduced me to the new building's architect, and we went through what I wanted in the construction of the studio and the acoustic treatment. This was a two story building with a penthouse on the roof. The second floor was leased out to another firm during construction, but later housed the camera room which contained the animation cameras, and other spaces for the animators and script people. The corporate offices were on the first floor at the rear of the building. There was also a concrete and steel stairwell closed off by fire doors at the top and bottom. Bob and I wired it up to use it as an echo chamber and it sounded very good. We later had to install a do not enter light to keep people from using the stairs during recording. Likewise I worked with the electrical contractor to work out the placement of electrical conduits from the power vault for the three phase feeds to the film transports and the projectors (which I ordered locally from Century Projector Co.) I found a company that made large digital readouts that could be used as footage counters for the large screen in the studio, and made a smaller version myself for the control room.

Bob and I unpacked and installed all of the equipment in the machine room which was immediately behind the control room. The projectors were installed by a local film supply house. The Century was for 35mm, and had a composite motor that could run stand alone or locked to the distributor system that ran the transports in the machine room. The distributor itself was located in the basement in the power vault. For 16 mm proof prints we had a 16 mm Eastman Model 35 equipped with a TV pull-down. It too, could be run on the distributor. When it was finished, Bill Hanna asked me to join them as Chief Engineer. I thought about it, but respectfully declined after learning that the local IATSE union would not honor my transfer from Chicago. I think that I probably would have declined anyway out of loyalty to Bill and United.

After marrying his third wife Miriam, Bill bought a home in the Hollywood Hills not far from the famous HOLLYWOOD sign. Bill had turned some space under the rear of the house into an electronics lab and workshop. I spent quite a bit of time there converting some of our old Telefunken and Teladi power supplies to work with some of the newer condenser microphones we were getting. It saved us from having to move power supplies from building to building each time we moved the mics, which was often. In those days, the condenser microphones all had tubes with high voltage power supplies.

**Bob:** Jerry and I thought a great deal of Tony Parri. Outside the area of Bill's engineering genius, hiring Tony was one of Bill's most inspired decisions. Tony would make apparently off-the-wall comments, which I later realized was his way of directing people to a better choice.

**Jerry**: As Controller, he held the company purse strings with great wisdom. He never minced words and would tell you exactly what he thought of your idea. If you had thought it out and presented it clearly, he was quick with praise and helpful suggestions although you still might not get the funding you were seeking. He always had a cheerful greeting when passing in the office hallway.

I went to Tony and asked for more money to buy some additional film equipment and have a cabinet made for my latest creation. I hand built a 35 mm system which projected into a TV camera (black and white in those days) We then bought a portable Ampex 1" video recorder and mounted it in one of our roll-around cabinets. The idea was to have the film dropped off the afternoon before the scoring session, and transfer the film onto videotape, which was played in the studio on a couple of 21 inch black and white monitors placed strategically where those who needed to see the picture could see them.

We didn't have to use music stand lights or dim the room lights, and we could record a monaural version on a track of the videotape, so we could play back the take for the conductor almost instantaneously. It was a big hit, and we used it almost every week. We even did live recordings of scores such as were used on the Gunsmoke TV series. The projector was eventually installed in the old dubbing room at Western along with the Stancil-Hoffman that was joined by a couple of Magnatech machines and another new "distributor" for an interlock system at Western.

The Distributor was placed in the back storeroom. This allowed us to view the film "live" from the projector to the studio next door. One Sunday morning Frank, Bob Felthousen and I strung a coax cable across the Columbia Pictures buildings and a side street to feed video to United. We were in the film business!

UA saw the need for a special piece of equipment to generate the click track that was related to the number of film frames. Up until this time the clicks were recorded on magnetic film and edited by film editors to fit a particular scene. The clicks or pops were generated by punching a hole in a piece of unexposed film leader and passing it through the optical head of a projector. In any case an unwieldy but necessary system.

Bud Morris' son, Roger is a talented digital system designer. For UA, he developed a digitally controlled click generator that was locked to the 60 cycle line and was related to frames of film. They called it a Digital Metronome, and it had thumbwheel switches on the front to select the frame rate. It had line output levels and could be patched into any headphone distribution system. There's probably not a studio in Hollywood that didn't eventually have at least one.

**Jerry:** It seems to me that there was some stir over trying our hand at consumer products at UA. As I recall it started out as a small portable stereo phonograph. I don't think it ever got past the prototype stage. Bill was still tinkering with the UA 175 limiter he had designed, and trying to get together a new modular input module to replace the old 610 unit. By this time the UA facilities at Western were getting tight, and Bill and Bud Morris decided to move the manufacturing, development, lab and offices to a larger building in North Hollywood. UA bought the name and assets of a company called Teletronix (known for their limiting amplifier of that name) and shipped all of the parts and equipment from that factory to ours. UA also acquired the proprietary designs and assets of Waveforms, Inc. of New York, a manufacturer of precision audio laboratory equipment and measuring devices. Alan Byers, formerly president of Waveforms was appointed product manager for the newly formed instrumentation division. UA also started to manufacture an updated version of the Teletronix Limiter about that time, This is when UA became a full-fledged audio brand.

About this time I started a company of my own on the side, called Studio Engineering Consultants. At first it was just to facilitate my work with Hanna, Barbera, but later I expanded into other areas. I still maintain the Company name today, but I no longer rep equipment companies.

A small company near San Diego called Auratone had started to gain the attention of studios around the area with a small bookshelf speaker and I purchased a pair to see what all the fuss was about. The speakers were quite amazing, and our clients were requesting their use for mix-downs as they felt they sounded just like the car radios of the day and so they wanted to mix their singles on these speakers so that they would know what the music sounded like in a car. The speakers used a 4" Quam speaker in an essentially sealed infinite baffle. Simple but very effective. I became a dealer for them in our immediate (Los Angeles) area.

I met a fellow at the studio who was in the importing business and had some great contacts in Japan. We got to talking and he said he thought we could make the little speakers ourselves. He asked me to provide him with one of the Quam drivers which he sent to Japan for prototyping (copying). We received and tested about four or five of their drivers in Auratone cabinets in our home built anechoic chamber until we were satisfied that we had a good match. Then my friend had some cabinets made by a company in the San Fernando Valley, ordered a 100 drivers from Japan and we started pushing them. They were packaged 2 to a carton, and they were an immediate hit. I must have sold several hundred pairs to the studios around town under my "SEC" label.

My friend was importing a pair of headphones that had a very nice sound and were being used by some of the studios for overdubbing and general cue use, We got to talking about the fact that we neened so many of them during film scoring, as all the musicians had to have a pair, but most of them wore them with one side off of their ear so they could here their instruments. I asked if he could have the headsets modified. He said maybe it would be possible if the numbers were right. I went out on a limb and said I could make the numbers right if he could make the price right, and make the headsets the way I wanted them. This was my first foray into having something custom made for me, and I was excited. I knew that film studios all over Los Angeles were paying big money for war surplus military headsets like the ones that were built into airmen's helmets and Tank drivers helmets, and the supply was getting very short.

I ordered a headset with a straight 20 foot *non-coiled* cord with standard phone plug. Just one earphone with a nice comfortable vinyl cushion and a soft over the head headpiece with a soft rubber end opposite the headphone, The impedance was to be 1000 ohms so that a large number could be plugged into the low impedance output of a standard cue amplifier without loading down the line. This meant that each headset had to contain a bridging transformer, which offset the price of the missing earphone, bringing the price about in line with their standard double headset. I ordered 100 units to start with, and sold most of them to United/Western to replace our old low impedance phones. When the word of mouth got out from the musicians using these headsets in our studios, I began to get orders from other big film studios around town. I sold several hundred of these headsets (under the SEC label) to many studios, and had repeat orders from many of them. They were ideal for use with the UA Electronic Digital Metronome which was now in use all around town. Then according to our agreement my importing friend began selling the headset under his own name to several Electronic Parts Distributers in the area and around the country. We did quite well with those. I decided to back off from merchandising as it was starting to take up too much of my time, I still did, and still do a little consulting from time to time both in audio and video surveillance systems ( a holdover from my years at ABC), I was always careful to avoid any conflict of interest with my job at United/Western, and Bill didn't seem to feel that there was a problem, as many of the people I dealt with were regular clients of our Companies and it was encouraging business.

With the area at Western vacated, the tape library was moved into the area where the UA assembly area had been, and I moved the maintenance shop into UA's old design area. Reprise Records was looking for some engineering office space, and moved into the old UA front offices. We also built a mix-down room for them in one of the rooms near the front lobby. Reprise hired Lee Herschberg from United to run that room and do mixing on their sessions, most of which were done at United or Western.

**Jerry:** In the Fall, Columbia booked Western Studio 1 to do a scoring date for a Western movie (I don't remember the name of it) with a rather large orchestra. When the day came to record, the assigned engineer called in with a family emergency and was not able to come to work. I was assigned to do the job. I didn't have much time to think about it and immediately began to lay out a studio setup for the setup techs to follow. This was a 32 piece orchestra, and all seats had to have headphones for a "click" track. We would be using our proprietary UA "Digital Metronome" tempo generator to feed the cue amplifiers. While I had mixed many smaller sessions for film (mainly for cartoons), I had never done one this large in this room. I was way over my head, so I went back and looked at some of the setup sheets for past sessions and came up with something workable. At session time the Columbia Director walked in, and not seeing his regular engineer nor knowing the one he saw, was not pleased. Accordingly the session got off to a bad start. Fortunately the musician's contractor was a very nice lady I had worked with on many occasions and she spotted the tension brewing immediately. She stepped right in to soothe the director, assuring him that I was capable of doing the job, and then while he was out in the studio, she told me not to make her look bad, and proceeded to make suggestions as to how to get what the director was looking for. Since most of my mixing was pop material I naturally was going for an upbeat pop sound, but the Director was looking for a heavier more somber background sound. Once we were all on the same page, everything fell into place and the session went smoothly after that.

**UNIVERSAL AUDIO, North Hollywood**
**1965 (still)**

**Bob:** I moved to North Hollywood with Warren Gilman 'Gil,' Jane, and Paul Jeansonne, as I had become very interested in console design. Gil was doing console design, and supervising quality control. Jane Geier was draftsman and mechanical designer. She was an alumnus of Cinema Engineering, which had been a grand old name in professional audio. Paul Jeansonne was production manager. As Jerry has explained, UA was moved to North Hollywood, and I moved with them. My title was "Sales Engineer," but I did a little bit of everything as the workload required.

Gil was a very interesting person who had no patience with stupidity or people who didn't think for themselves. If I came to him with a question carefully thought out in advance, he would respond by asking me several questions, each of which I would answer. After my last answer, he would cock his head, then look at me as if to say, "Well?" I realized then I had the answer to my original question. He made you think!

He would even use the same technique with Bill, though not in as pointed a fashion.

Deane Jensen (yes, the same person of transformer fame!) came to work for us fresh from the East Coast. Actually, he came along with somebody else who was applying for a test technician's position. That person was turned down, but after talking with Deane, Gil hired him.

**Bob:** At this time, we were manufacturing the 175A limiter, which was a 3-½ inch rack unit, with not much space inside. Suffice it to say, the limiter used vacuum tubes with a B+ voltage of around 300 volts. If I was walking through the lab, and Deane was measuring voltages inside the unit, I would loudly clap my hands to simulate a capacitor popping. Invariably, Deane would jump! Fortunately, he never injured himself, nor a 175 under test.

Several years later, when I was running my own company, Deane was working for us on new product development. One afternoon after work, we were looking at an AM-FM tuner that Bill Brandt had brought in to test. Bill was our systems design engineer. I was leaning over the tuner, which wasn't even plugged in, and Deane had his chance. He loudly clapped his hands, once. I jumped, just as he did, then Deane broke up. His comment was to the effect that he finally got even!

Back to Universal Audio; I got very involved in console design, under Gil's tutelage. We built several boards for local companies. Bill wanted to use the name "Studio Electronics Corporation" for the custom boards.

**1966**

Bud Morris and I landed a contract to install sound systems at the Aladdin Hotel in Las Vegas. The hotel was adding a full casino and a showroom. I was manufacturer's representative, and had two electricians working for me. One of them was a working foreman, so I was able to talk directly to him.

For part of the showroom system, we had an intercom system using W.E. Co. headsets and a DC power supply. The electricians finished wiring the system, but it didn't work. I called Gil for help, but he was occupied with other work, so he was unable to help me. I called the vendor in Los Angeles, they suggested a choke in the DC line. Suddenly the light came on! The DC power supply looked like a short to the audio. I ordered a choke from them, we received it on the jobsite the next day. I had it installed, then the system worked! Never overlook the obvious!

Caesars Palace was being built, and Ed White put in some good words for us. During the construction of Caesars Palace, the owners were concerned about sound isolation between hotel rooms. They asked us if we would perform the measurements. Bill brought me up to speed on the techniques, then Bud and I went to Las Vegas on one of our sales trips.

I did the measurements in the late afternoon, after construction had stopped for the day. From the EC (Electrical Contractor), I borrowed a power cable which was used during construction. I went up to a pair of rooms where two doors had been installed for my measurements. As I recall, I borrowed a small tape machine, a power amplifier, and a speaker from

URCON, our studio in Las Vegas. And, of course, a 7-inch tape on which I had recorded pink noise.

The walls were finished, but no carpeting. I found some discarded fiberglass insulation that I stuffed between the bottom of the doors and the rough flooring. I set up the tape machine and speaker in one room, set the noise level in that room sufficiently high to get a useful measurement. I then stuffed the insulation below the first door, went into the other room, stuffed insulation below that door, then measured the noise level.

It took several tries before I got the levels right, for if the sound level from the speaker was too high, I would have distortion and possible speaker damage. If the level was too low, I wouldn't be able to measure the difference between the ambient and the desired noise. I finished the measurements, discussed them with Bud and Bill, then gave the results to the owners. Suffice to say, their construction design and methods were excellent; they had no problems.

After much discussion, and several trips that Bud Morris and I made to Las Vegas, we got the contract for the main showroom sound system, lounge sound reinforcement system, and public area background music systems. It was decided to use the design for a board we had built for the Desert Inn for the Caesars Palace board. This simplified system design and especially shortened construction time.

Bill and Bud nominated me to install the systems. Because of our ties with IBEW in Los Angeles, and because we were hiring sound technicians from the local in Las Vegas through the electrical contractor, the local allowed me to be a non-working foreman, rather than just a manufacturer's representative. I was on the jobsite at 7 AM, and was busy with directing my three-man crew, and tracking the status of various areas where we would be working next. I had to remember not to reach for a tool to show the techs what I wanted done. I was a non-working foreman! I worked there each week, and went home for weekends.

The light and sound booth was across the back of the showroom, which was called the Circus Maximus. The console was in the sound booth, and a two-inch conduit went from behind the console, up and over the show room ceiling, then down to a junction box at stage left. As I recall, we had about 37 mike cables to run. With one of the sound techs, I counted conduit lengths, then estimated makeup lengths at each end. The cable was Belden 8761, and I had requested enough cable so the techs could place wire markers, then make one pull without having to measure the cable length. The sound tech thought we should add about 25 feet to the run, just to be sure. I was confident of our measurements, and told him to leave the length as we calculated.

I left the booth, then came back just before lunch. One of the other sound techs was in the booth. I saw the mike cables across the floor, with about 25 feet looped around. I asked one of the other sound techs if they were done. He quietly said he would tell me about it after lunch.

Later he told me that the other tech thought they should add 25 feet to the run, which they did. When they finished pulling the cables, they had 25 feet left over! After that incident, I never got an argument from the sound techs.

There were two or three light troughs across the showroom ceiling. The lighting instruments were to be hung by the electricians. On the drawings, we had shown three large speaker cabinets at specific locations in one of the light troughs. These were the main clusters to provide sound to the audience. I had become friends with the general foreman for the electric contractor, Gene Ray (I think was his name) but when it came down to business, we were on separate sides of the fence.

We were discussing where our clusters would be located, and Gene said they had to be at a different location, as they would interfere with the lighting instruments he was going to hang. Gene was a large individual, who participated in rodeos when he wasn't working. We got into a spirited argument that was finally resolved when the consultant from the lighting company told Gene that these locations were only arbitrary, as the light techs to be with Caesars Palace would rehang the instruments as required for a particular show. I thanked the consultant for his clarifications and comments. Several years later the consultant and I met in Los Angeles on another job, and we reminisced about that incident.

Caesars Palace opened in August 1966. Even though they had hired a soundman, Dave Rogers, they asked me to be there in case of problems. Andy Williams was the opening star, backed by a full orchestra. The only power on the bandstand was used by the electric guitar during rehearsals. Somebody with Andy Williams' entourage plugged in his condenser mike power supply to the one receptacle. When the show started, the guitar player realized his amp was off, so he unplugged the cable that was using his receptacle. This was, of course, Andy Williams' mike. It went dead, Andy looked at it, tossed it over his shoulder. Somebody brought him a replacement mike, and the show continued. After the show, Dave and I checked the mike. It still worked, but didn't sound too good.

Through a friend in show business in Las Vegas, I got a rare treat. Since my friend knew the stage manager at the Stardust, I asked my friend if he could arrange for me to be backstage during a show. After talking with the stage manager, my friend told me where to be and what time.

Long before show time, I met the stage manager. After talking with me, he seemed to decide that I would behave myself. He advised me that when he put me in a particular spot, I shouldn't move from that spot, otherwise I would endanger the stagehands and myself.

We went onstage, dimly lit by overhead lights, and the stagehands were moving sets around. He put me in a location, again advised me not to move until he came to get me. I watched the very organized confusion, then as the show started, he took me to one side of the stage. I watched the show, and the stagehands quietly moving about, getting ready for the next act. As I remember, the opening act was a dance number, but done in a large water tank onstage with a waterfall, and two dancers performing in the water and the waterfall.

Since the show had been thoroughly rehearsed, and performed a number of times, the stage manager's main task was watching to see if there were any problems. That's why he had time to lead me around. After the number finished, the stage manager led me near the water tank, which I realized was on a large elevator. The elevator quietly descended to the lower level. We got off the elevator, and several stagehands disconnected the large hoses from the waterfall assembly, then pushed the water tank upstage! It was on large casters for mobility.

I watched the rest of the show from various locations, and after the show finished, thanked the stage manager, and went back to my motel. Only because we were installing the sound systems at Caesars Palace, was I able to do something that not many people have done. I was very impressed with the professionalism of the entire crew at the Stardust, which I'm sure is typical of the various shows in Las Vegas. I thanked my friend the next night.

**Jerry:** About this time, a number of new people were hired at Universal Audio, among which were Brad Plunkett (a bright young man from Lyon & Healy's Hammond Division), Juergen Wahl, and a few people to run the office and purchasing. They hired a shop foreman (Dick Simisky) to manage the assembly line of ten girls. Also added was a shipping clerk. The name UA was changed to UREI (United Recording Electronic Industries). I got out there only occasionally (when I needed something or to pick someone's brains for ideas).

We also hired a number of people at the studios to help with the maintenance and upkeep of the equipment, and to help with setting up studios. We were very busy during this time and we were keeping the studios humming. I hired Don Foster a maintenance engineer formerly of CBS who was once a sound effects engineer for many radio serials. He was also the inventor of the electronic gunshot used on many Western and Cops & Robbers radio shows. He was a kindly gentleman and took great care of our mastering lathes and systems, checking them for frequency response and general health each morning.

After Bill had set up the shop under his house, he and I spent a lot of time together working on a number of projects, one of which was a reverberation unit that used an electrostatic charge on a rotating metal disk. The charge was laid down and picked up every rotation of the disk, and was based on an invention used by Fender Guitars for reverb in their instrument amplifiers. Bill was elaborating on the principle with his own version, and much experimentation was done at his home. It became the UA Model 120 Electrostatic Reverberation Simulator. I believe there weren't more than about ten or fifteen units ever built, but there are some out there somewhere. I believe at least two were installed at Ray Charles' Tangerine Studios in Los Angeles.

**Bob**: After Caesars Palace, we built a board using the 610 Module, which had been made famous in Studio 3 at Western. This board was installed at Tangerine Records' studio that Ray Charles, the blues singer, had built in Los Angeles. That was also the first sale of the Model 120 Electrostatic Reverberation Simulator. I installed the system, and watched while their accepting engineer, Tom Dowd from Atlantic Records, showed Ray how to use the board. I realized then how

Ray's memory, hearing and sense of touch made up for his blindness. I had worked for Tom in New York, before my wife and I had moved to Los Angeles.

**Jerry:** Around this time 3M came out with an 8 track version of their new transport. It somewhat resembled the Ampex cabinet in that it had 4 amplifiers with integrated meters above the transport and 4 more below the transport. They worked quite well, and we ultimately bought four, placing two in each building.

**Bob:** We then built a large board, which used Bill's new 1108 solid state mike preamp, again with his cascode design. That board was for Liberty Records. Deane and I worked on that, again under Gil's tutelage. I learned a thorough lesson while working with Deane on the console checkout. When two people are working as a team, at any given time, one and only one must be the leader. That role can shift back and forth, depending on a number of factors, but only one person can call the shots.

What happened? Since the solid state design was new, as were the power supplies, we were proceeding cautiously. We were investigating power supply noise and DC power system corruption. The supplies, made by Lambda, had provision for remote sensing. This meant that the power supply would regulate according to the load at the console, rather than the power supply terminals. Among other things, it automatically adjusted for power cable length.

Deane and I violated the team rule. He disconnected the sense leads at the same time I powered up the power supplies. The supplies went temporarily mad, fortunately not destroying the 1108 amplifiers. Deane spent some time in repairing the supplies. I didn't see Bill much during that time, as he was working on new product development.

### 1967

After the Liberty board, Bill and Bud decided the custom console business was too labor-intensive, and decided to stop manufacture of custom boards. Not all was lost for me, as Bill asked me if I wanted to go into business for myself. They would give me a Universal Audio dealership, and give help as they could. So I started Bushnell Electronics Corporation. He also turned over a contract for a custom sound reinforcement board for a company in San Francisco, who was happy to deal with me. That was in the fall.

### 1968

**Bob** continues: About a year later, United needed a mix-down board for Reprise Records. Jerry contacted me. Jerry and I jointly designed the board, and my company built the board. I made one mistake that later caused Jerry some headaches. We had two wiring ladies, very capable people indeed. One of them asked for clarification for wiring TRS (tip, ring and sleeve) jacks. I showed her the connections. Unfortunately, I transposed the tip and ring, which were normally the high and low sides. Therefore, all the jacks in the board were wired with reverse polarity.

### 1969

Bill decided to replace several boards at United, Western, United in Las Vegas, and United in San Francisco. He asked me if we would like to build them, and on a cost-plus basis. We were delighted with the contract, and with Bill's giving us further business. We built five boards for Bill, all at a reasonable profit.

The boards were designed around a new module that Bill's R&D lab had designed. Bill had hired John Jarvis from Langevin to help design the new module and to this end he opened a small research facility in a storefront in Tarzana. He spent a lot of his time there and wasn't spending nearly as much time in the studio as he had been previously. Bill didn't do the design himself, but he might have been in closer touch with the project. The module was in the late stages of development, and I was regularly hounding Bob Doherty for more details. We finally got the first run of modules. Warren Dace, our chief engineer, and I plugged them in and turned on power. The R&D lab's efforts to simplify the design had caused some problems. I won't go into details, but it took concerted effort on everyone's part to correct the problems.

**Jerry:** Bob Doherty decided to go into business for himself, and started a tape duplicating business in the Western building. He later moved it to Burbank. Bill then hired Don Sears, owner of Sound Recorders of Omaha who had a good track record and well-known studio in the Midwest. He became General Manager of United and a good friend. He was there until 1971, when he left to return to his studio in Omaha.

We liked the 3M recorders so well that when they offered a 2" 16 track model we bought two of them. These reverted back to the circuit card style used earlier in their two track machines, except each channel had its own single card which contained all of the electronics for one channel except for control circuitry. 3M wisely created a separate remote control console that contained the controls for each channel and included transport controls also. This allowed one to place the remote next to the mixer or recordist so the machine could be placed out of the way. It was a smooth running transport.

At that time some of our tape machine connectors were causing problems, which prompted us to do away with the unwieldy W.E. Company 241 plugs that we had been using on all our tape machine inputs and outputs. They were handy during the early days of stereo when we often got tapes that had out of phase tracks, as we could simply turn over the plug. But with the advent of machines with more than four channels, came huge wads of plugs and cables, and it was time for a change. Don suggested we use Amphenol multi conductor military style connectors that had 56 compact gold wiping contacts, which he had used successfully at his studio. We found that they worked well, but were susceptible to breakage when dropped, which happened all too frequently. We devised a metal surround shell for the connector, and that solved the problem.

Of course, this triggered a mass renovation project because now we had to convert six or seven studios over to the new connectors and each studio had at least three machine positions. In order to expedite the changeover, we had to do as much pre-preparation of new wall boxes and covers with cut-outs for the new 56 pin receptacles as possible. Because multi- track machines were moved around from studio to studio, this had to be done very quickly to maintain the usability of the studios. We took one long weekend to do everything. One team wired the studios at Western and another did the studios at United. We had made up the machine cables ahead of time, so they only needed to be plugged into the machines and dressed out. For these cables we used 19 pair multiple cables with 3 spare pairs per cable. The cable was splayed out three feet from the machine plug end and each pair was covered with clear tubing over a numbered wire marker. We used shrink tubing where the pairs broke out of the cable. The other end had the 56 pin multi-conductor plug. It took almost two weeks of intermittent work to make up enough cables for all our multi-track machines.

## 1970

**Jerry:** Tony Parri was suffering from a number of ailments, among them cancer. He had been very ill for a number of months, and passed away in June of 1970. It was a very sad time for the Company. I truly missed his smiling greetings in the hallways.

**Bob**: At Tony's funeral I recall Bill quietly crying and telling me that he was going to miss Tony. I think that Tony's financial acumen matched Bill's engineering acumen. They were very close.

## 1971

**Jerry**: In the Spring of 1971, Bill approached me about doing some lecturing on recording subjects for his friends at Brigham Young University Extension School's Annual Audio/Recording Seminar. I spoke there for two years in a row doing lectures on "Stereo Mastering and Groove Geometry" and "Studio Maintenance Procedures for Multi-Track Recording Equipment." He also talked me into being the Equipment Chairman for a Los Angeles Chapter AES convention. I didn't know it at the time, but it was invaluable experience. I served in that capacity for two years.

I also presented a few papers at the AES convention on behalf of Universal Audio and myself. One in particular was very well received; it was on the use of video to aid in scoring music for motion pictures and TV productions.

**Bob**: As a side comment, I also lectured at the BYU seminar, but earlier, in 1969. My topic was "Console Design and Specifications."

As Jerry did, I presented a paper at an AES convention. The topic was the 176 limiter, which Bill had designed to provide various compression ratios.

**1973**

My company was in financial trouble. I had forgotten that profitability is required, as well as quality of our boards. Shortly before we closed up, Bill, Bud Morris, their controller and their financial adviser spent an entire evening with me at our factory. He recommended a lawyer who was experienced in bankruptcies. Bill expressed the thought at the end of the meeting that if I had come to see him some six months before, they might have been able to do something. As it was, Bill and the bankruptcy lawyer were of invaluable help in helping me close the business. I'm still grateful to Bill for that, among many other things through the years.

**Jerry:** Bill had been divorced from Belinda (who had not been active in United's studios after they moved from Chicago, other than some editing sessions). She also worked as Tape Librarian for Liberty records for a time. She died shortly after from liver complications. Then Bill had a severe heart attack (he smoked heavily and drank some). After recuperating from multiple bypass surgery, he married Miriam Simons, his third wife, who had been a secretary at Reprise Records and was introduced to Bill by Sinatra at one of his recording sessions.

By this time, I had been named Vice President of Engineering for the recording companies, and when Don Sears left, Bill turned the studio operation over to Jerry and Joan Barnes from Texas who ran it for several years. Bill was spending much less time on studio matters and was concentrating on UREI, the new name of the manufacturing branch. They were marketing a modified version of the Altec 605 system in a proprietary speaker box called the UREI Studio Monitor. The idea was created by Ed Long, who had received a patent license from Jet Propulsion Labs in Pasadena. At JPL, Dick Heyser had developed and invented the Time Align system, which led to some major changes in loudspeaker designs among all manufacturers.

Also, Bill was experimenting with a system for using a long length of coiled tubing with special transducers as an alternative to the metal plates used for reverberation. The original idea was patented by Duane Cooper, a physics professor at the University of Illinois.

Jerry and Joan Barnes were trying to make the studios work for the smaller groups, as sessions with large orchestras were becoming fairly rare. Also most groups were using their own mixers who always knew more than anyone else about how to mix their groups or set up studios for them, although most knew only the basics about the equipment itself.

By this time, we were up to 24 tracks, and with each step up you could hear the engineers saying "What do we need more tracks for? What are we going to put on them?" But the clients were asking for them and if we didn't give it to them, one of our competitors would. So we got two machines, and it wasn't long before the engineers were finding ways to sync them up to get 46 tracks (one track of each was used for syncing). I never did see the sense of it except it allowed the mixer to edit between takes on the fly while mixing down to 2-tracks. Since none of the boards were equipped for more than 32 tracks at the most, one had to do some tricky patching on the fly also. Neither United or Western had any computerized mixing consoles while I was there.

So the studios were becoming basically rental units requiring hand holding by the maintenance people who had to cater to the whims of guest mixers. By this time, the heyday of large studio complexes had pretty well peaked and we were headed into a slump from which the larger independent studios would never emerge. Their operating overhead was just too high to be supported by the flagging market, and larger studios were being subdivided into smaller rooms equipped with the basics for recording. Many musicians and groups had built their own small studios and were providing record companies with finished product ready to master.

**1974**

**Jerry continues:** Most of the top executives of our company had retired or been laid off, and some of the mixers on staff were doing mainly agency work and many had left the company. I saw the handwriting on the wall, and in 1974 Bill and I had a long talk, and he told me that it was going to be impossible to keep me on much longer, and he advised me to keep my eyes open for other opportunities.

At this point I have to say that in all this time (seventeen years), Bill had been like a second father to me. He was there to talk to if I had a problem and he and his family treated my family like we were part of theirs. Our children played together when they were small and we were all very close. Bill's wife Miriam succumbed to cancer, leaving him with two teenage sons to care for.

So it was not without some painful moments when I said goodbye to United and its companies, and especially to Bill who had been my mentor and good friend for so long. It would be impossible to list the tremendous amount of knowledge he imparted over all those years together and he always encouraged me to try new things. I accepted a job as Chief Engineer with ABC Recording Studios Inc., which had been courting me for some time. I spent another twenty-five years at ABC. I saw Bill about once a year at the AES conventions that he attended as an exhibitor with UREI. The meetings were cordial, but we had lost that spark of camaraderie.

Bill finally sold UREI to Harman Industries in Van Nuys, which also owned a number of other electronic companies including JBL. He married his fourth wife, Caroline and lived with his new family in an area that he had always been fond of, Ventura Keys, CA. Bill loved to sail and he had bought a yawl sailboat some years earlier. Now he had it moored in the channel right behind his home.

Bill passed away on April 13, 1989. Bob managed to find me (I was visiting in Chicago at the time) and I took a flight the next day. Bob and I attended Bill's funeral service on the grass at Valley Oaks Memorial Park in Westlake Village, CA. There were a number of highly regarded musicians who had gathered together a band to play some of Bill's favorite jazz tunes, and there were many record executives there as well as colleagues. He is missed and will be remembered for his innovations to the world of recording and his inventive style of mixing, copied by so many.

He is survived by his fourth wife Caroline, his daughter Sue and his three sons, Scott, William (Bill), and Jim. Bill has followed in his footsteps and has restored the Universal Audio name. He is manufacturing improved versions of the UA 1176 Limiter and many other UA products. We wish him great success in his venture. I'm sure his Dad would be proud.

**Many Thanks to All:**

In addition to Bill and his sons, we thank all the people we have worked with. They are a large part of this small book. Unfortunately, some of them have passed on. Some still live in Los Angeles, some we have lost track of. We were able to reminisce with Bud Morris, Bill Perkins, Manny Berman, and Andy Richardson at Bill's funeral.

Between Chicago, Los Angeles, Las Vegas, and San Francisco, this is a fair collection of friends and people. If we have left anyone out, our apologies. These are the people that we remember:

Legend as to the various studios and companies:
ABC= ABC Records
Atlantic = Atlantic Records, New York
BEC = Bushnell Electronics Corp., Los Angeles
MR = Master Recorders
UA = Universal Audio, Chicago
URC = United Recording Corporation, Los Angeles
URC-C = Universal Recording Corporation, Chicago
URCON = United Recording of Nevada, Las Vegas
URC-SF = United Recording of San Francisco
UREI = United Recording Electronics Industries, formerly Universal Audio
WBR = Warner Bros. Records
WR = Western Recorders, later purchased by United
WHITE = Stan White Loudspeakers, Chicago

| | |
|---|---|
| Jerry Barnes | URC |
| Joan Barnes | URC |
| Artie Becker | URC (deceased) |
| Betty Belson | URC |
| Manny Berman | CPA for URC-C and URC |

| | |
|---|---|
| Remo Biondi | URC-C |
| Don Blake | WR, URC |
| Eddie Brackett | URC |
| Bill Brandt | BEC |
| Chuck Britz | WR, URC |
| John Boyd | URC |
| Al Butow | URC |
| Alan Byers | Waveforms, UREI |
| Malcolm Chisholm | URC-C (deceased) |
| Bernie Clapper | URC-C (deceased) |
| Ray Combs | UREI |
| Mason Coppinger | URC-C |
| Warren Dace | URC |
| Bowen David | URC |
| Frank deMedio | URC |
| Lina diFiore | URC |
| Phil Diamond | URC |
| Bob Doherty | URC |
| Tom Dowd | Atlantic |
| Jim Economides | URC |
| Alan Emig | UREI |
| Jack Eglash | URCON |
| Curt Esser | Architect for URC-C |
| Bob Felthousen | URC, WR |
| Don Foster | URC (deceased) |
| Lowell Frank | WBR |
| Dorothy Friend | URC |
| John Gaines | URC |
| Jane Geier | UREI |
| Don Geis | URC, URC-SF (deceased) |
| Mary Geis | Don Geis's wife |
| Warren Gilman | URC, UREI |
| Bob Golden | URC |
| Leila Greenstone | URC |
| Hal Halvorsen | URC (deceased) |
| Bobby Hata | URC |

| | |
|---|---|
| Dave Harrelson | BEC |
| Eileen Harrelson | UREI, BEC |
| Dale Harrington | URC-C |
| Wally Heider | URC (deceased) |
| Shelley Herman | UREI |
| Lee Herschberg | URC, WBR |
| Rudy Hill | URC, WBR |
| Bones Howe | URC, WR |
| John Jarvis | UREI |
| Paul Jeansonne | UREI |
| Deane Jensen | URC, BEC (deceased) |
| Jennifer Johnson | URC |
| Alice Jones | URC |
| Ben Jordan | URC, WR |
| Phil Kaye | URC, WR, ABC |
| Larry Kissner | UA |
| Bill Kirkpatrick | WR |
| Dorothy Lee | URC |
| Henry Lewy | URC |
| Lanky Linstrot | URC |
| Jimmy Lockert | URC, ABC |
| Ray McKinnon | URC |
| Sandy McNeely | UREI |
| Al McPherson | URC |
| Brent Maher | URCON |
| Bud Morris | URC, UREI (deceased) |
| Ethel Morris | Bud Morris's wife (deceased) |
| Roger Morris | UREI |
| Ken Stone | Consultant for UREI |
| Dick Stover | UREI |
| Balasz Nagy | UREI |
| John Neal | URC |
| Alyce Ordman | BEC |
| Mildred Parri | Tony's wife |
| Tony Parri | URC (deceased) |
| Tom Parrish | URC-C |

| | |
|---|---|
| Walt Payne | URCON, URC-SF (deceased) |
| Bill Perkins | URC |
| Brad Plunkett | UREI |
| Bill Porter | URCON |
| Lloyd Pratt | URC |
| Miriam (Tooky) | Putnam (Bill's 3rd wife) (deceased) |
| Bill Putnam | Bill and Miriam's son |
| Jim Putnam | Bill and Miriam's son |
| Len Reightley | URC |
| Andy Richardson | URC |
| Belinda Richmond | (Bill's 2nd wife) URC-C, URC(dec,) |
| Frank Richter | URC-C |
| Lenny Roberts | URC |
| Ted Robinson | WR, URC (deceased) |
| Bunny Robyn | MR, URC |
| Roy Rogers | UA-C (deceased) |
| Phill Sawyer | URC (deceased) |
| Nook Schreier | (David Carroll) (deceased) Mercury Records, Chicago |
| Lillian Sewell | URC |
| Dick Sexty | URC, WR |
| Mike Shields (Nemo) | URC |
| Joe Sidore | URC |
| Dick Simisky | UREI |
| Alan Solomon | URC |
| Bill Stoddard | URC-C, URC |
| Mel Tanner | URCON, URC-SF |
| Dick Tullis | WHITE, URC-C |
| Tee Jay Vaughn | UREI |
| Bob Weber | URC-C |
| Stan White | Stan White Loudspeakers, Inc. |
| Carolyn Wolf | URC |
| Winston Wong | URC-SF, URC |
| Randy Wood | Dot Records, Ranwood Records |

## OUR VIEW OF THE RECORDING INDUSTRY

Let's briefly look at the style of recording and processing through the years. As part of his development of the flat phonograph record, Emil Berliner physically connected a large horn to the recording mechanism. Then the musicians and singers gathered in front of the horn according to pecking order. Not really in pecking order, for the quieter instruments and singers were nearest the horn, and percussion was way out in left field. These days the percussion is way out in front.

We shall refer to the Telephone Company as Ma Bell, as that august organization was referred to more than once. Ma Bell, in particular Bell Telephone Labs, developed electrical recording, the recording horns were discarded, and the band or orchestra was miked according to the desires of the producer, the engineer, and even the conductor. From the early twenties to the late forties, all recording was done in real time. What we called a few years back, "Direct to Disk." Through the 40's, that was the style, there was no choice. Some overdubbing was performed. It wasn't very easy; the original disk was played back, then copied to the new disk together with the new material. Unfortunately, the wax master wouldn't stand very many replays, and the loss of quality was discernible. In the late sixty's, certain artists were making direct-to-disk albums for the ultra-purist High Fidelity addicts. Mostly solo instruments with mikes taken directly to the cutterhead at 78 RPM. Usually pressed on pure vinyl.

Magnetic tape recording provided the second real change in recording techniques, electrical disk recording being the first. Wire recording came first (just barely), but then Jack Mullin brought back a Magnetophone from Germany after WWII, and the fun began. Magnetic tape allowed a viable two-step recording process, in that the original recording was on magnetic tape, then edited, then transferred to a lacquer master. An electronics distributor in Chicago in the early fifties sold equipment to broadcasters, including Magnecord PT-6 and Ampex 400 tape recorders.

**Bob:** When I first went to work for Bill in Chicago in late 1953, he had just gotten three Ampex 300 mono tape machines, which would handle a 10½-inch reel.

Remember the LP record? That was a development of Columbia Records. Remember the 45-rpm record? That was a development of RCA Victor, in competition with Columbia. For a time, RCA sold classical works on stacks of 45 rpm records. They weren't very practical. At the same time, Columbia sold singles cut in the LP style, but on 7 inch disks, the same size as 45-rpm disks, but with a small center hole. They weren't very practical, either. The record industry finally settled on 10 inch, then 12 inch LP records for classical and albums; and 7 inch 45 rpm records for singles. Nowadays, you'll find lots of used LPs for sale, but no new LPs. For that matter, there might be six LP mastering houses in the US. They cater mainly to DJs, Rap Bands, and Solo Artists seeking tonal purity. The majority of listeners are quite happy with CDs. (They're also easier to copy!)

From monaural tape to two-track (not binaural!), then 3 track, then 4 track, then 8 track, then 16 track, then 24 track, then some magnetic recording of digital information, then hard-disk recording.

The need for mixers, consoles and boards roughly paralleled the irregular changes in recording techniques, up to about ten years ago, when board complexity far surpassed the recording technique itself.

There wasn't any need for mixing consoles in the acoustic recording era; there wasn't anything to mix. Electrical disk recording provided the nucleus for mixing consoles, but to call them consoles is stretching reality. They were mixers, 4 to 8 inputs, no eq nor echo sends. All that came later. 8 to 12 inputs were the rule through the early 60's.

If you listen to 78's that were recorded before the Second World War, or up to 1950; you'll be listening to the actual performances with very little editing. There might have been some by playing one lacquer, then cutting to another. However, that would have been a cumbersome process. Not until magnetic sound recording was available in the early 50's, was editing a feasible task. If you are listening to CDs created from 78's, you still have a pretty good idea.

Bill Putnam's mono board in the early 50's had 8 inputs, echo sends, no eq, but a tape return pot. As far as we know, Bill had the first use of tape reverb. Hard to count how many hits were cut on that board, which was in Studio A at Universal Recording, but Pat Boone, Muddy Waters, Stan Kenton, Duke Ellington were there regularly.

Through the 60's, boards had 12 inputs, 3 or 4 output buses, built in-house or modified from such mixers as the Altec 250SU and 250T3. One notable variation to in-house design was Bill Putnam, who had built his own boards in Chicago. After building boards for the new United Recording studios in Los Angeles, which began operation in 1959, he proceeded to build boards (3 track of course) for the competition and anyone willing to buy. He called the console company Studio Electronics Corporation, and later changed the name to Universal Audio. Does the name ring a bell? Bill finally stopped building boards in 1967, Liberty Records being the last board produced. It had 32 inputs, and 8 output buses.

About 1967 Electrodyne started the ball rolling with stock consoles, working from a modular approach. Inputs up to 20, outputs up to 8, it was a buyer's choice. That was the first use of ACN's, or Active Combining Networks, which were summing amplifiers. Don McLaughlin and John Hall developed those first summing amplifiers, using Fairchild 709 opamps. Quad-Eight, eager to join the ball game, came out with a module suspiciously similar to Electrodyne's, then later redesigned the module. Electrodyne introduced other major ideas; in each input module (the module itself being a new idea) were equalizers, echo sends, bus assigns. 8 buses in 1967, 24 buses in 1982; quite a change.

In the 70's, the English manufacturers made their presence known in the US; Helios, Trident, Rupert Neve. Various American manufacturers added their consoles to the market; the console buyer had an extremely wide range of boards from which to choose. Custom boards were beginning to price themselves out of the market, because of that wide range of semi-stock boards available.

In response to the problems of monitoring multi-track recorders and effectively using the advantages of individual track recording, console design shifted from a separate monitor section to "in-line monitoring." Since inputs could be controlled and selected, artists and musicians began requesting headphone feeds from selected inputs.

Bear in mind, we're discussing consoles on the American scene. England and Europe were different stories. The BBC backed a lot of research and design by console manufacturers. Unquestionably, UK's contribution to console design was a major factor. Therein may lie the rub. Did the English designers overly complicate console design, and still continue to do so?

Westrex was bought by Quad-Eight, then QE was taken over by Mitsubishi Pro Audio Group (MPAG), locally directed by a loser. MPAG was shut down by the parent company because the loser spent money without looking for return on their investment (sounds familiar?). QE was taken over by front money from Pioneer, locally headed by Bill Windsor (who had been sales manager for QE, then MPAG) and funded by the previous Japanese importer of QE.

Westrex was then taken over by a small projector company in LA. Frank Pontius had headed Westrex in Burbank. Jack Leahy had headed up RCA Photophone in Burbank. They joined forces, and were very involved in the operation.

Western Electric began to explore sound reinforcement (Public Address) in the early 20's. They provided the system for the inaugural address of President Warren

G. Harding (remember the Teapot Dome scandal?) in March 1921. A tremendous amount of basic work was done by WE through the years. They were later exporting sound equipment throughout the world, and they called the operation Western Electric Export. You can see where the name Westrex came from.

The Justice Department and AT&T came to legal blows about 1949, maybe later. The upshot of the decision was that AT&T was to divest themselves of the audio side of their business. The famous 639 microphone, 604 and 733 loudspeaker manufacture was taken over by Altec-Lansing; the motion picture sound equipment and disk cutting systems were taken over by Litton Industries as a division of Litton.

Incidentally, after Westrex Litton started development of the 3A Stereophonic Disk Recorder system, they went to patent it. They couldn't, because a previous patent had been issued to Arthur C. Keller and Irad S. Rafuse, dated April 19, 1938, patent #2,114,471. To whom? Who else but Bell Telephone Laboratories.

So between Western Electric, then Westrex, a lot of groundwork was done, providing a solid base for the audio and sound reinforcement industries.

## SOME DISCUSSION ABOUT STUDIO LAYOUTS

**Bob:** Generally, studio layouts were at the discretion of the engineer doing the session. Some of the more experienced "setup" people were consulted as to what was done on an earlier session. Sometimes the engineer would ask the setup person to do the same setup as on an earlier session. Most of the engineers had specific preferences as to mikes and room layouts. They also had specific mike to console inputs. As I mentioned on page 22, Bill used the same input assignment for his sessions. Input 1 was always the vocal; after all, what was more important than the vocal?

We don't have space here to show the basic studio layouts, but descriptions may help. Studio A at 111 E. Ontario was at the back of our space within a large office building. The room was about 30 feet wide, about 50 feet long, and the ceiling height was about 20 to 25 feet. The control room, which was in the center of the long wall, jutted about 5 feet into the studio. The approximate room volume was about 24,000 cu. feet. These dimensions are very approximate, as this information is from memory and some photographs. As I discussed earlier, the console was monaural, and the monitor loudspeaker was to the left of the console.

The studio walls were a combination of three or four elements. On the back wall, opposite the control room, were perforated panels firred out over fiberglass batts. On that wall, and on the wall to the right were large flat frames, faced with burlap or a more durable material over fiberglass batts. Also on that right wall were acoustical panels, probably 12 inches square, firred out.

To the left side of the control room was a large shell, with polycylindrical baffles. The shell was on casters, but not moved very often. Bill had also designed and built a vocal booth, also movable. This was 1955, Bill was definitely ahead of the class in studio design!

The floor was asphalt tile over concrete. I haven't the faintest idea of the construction behind the walls, except the building we were in was a large office building.

For smaller dates, the rhythm section (piano, bass, drums, and guitar) was directly opposite the control room. Woodwinds were to the right of the control room, and brass was usually in the shell. The vocalist was in the booth. If there were strings on the date, I think Bill put the strings in the shell, and brass to the right.

For brass, Bill usually used a 77DX, and a 44BX for woodwinds. The bass mike was an older WE design, later made by Altec as the 639 and drums was usually a 77DX, just one! When the Telefunken (really a Neumann U47) appeared, Bill used them on strings or brass. Depending on the vocalist, the mikes varied. For Pat Boone, I recall a 77DX being used.

Around 1955, we got an EV 666 on trial from Lou Burroughs. It was a good mike for vocals, as I recall. Most of the mike booms were Starbird, a very solid design.

**Jerry:** Well, as you can imagine, Bill didn't change his modus operandi much after the move. The new studio on Walton Place could handle much larger groups, and did on many occasions. But here we had three studios to work with, and we kept them busy for the most part. As before, there were vocal booths with three padded walls and one open side which usually faced the orchestra. These booths were on casters so they could be placed wherever it was convenient.

Usually some type of capacitor microphone was used with a cardioid pattern facing into the booth. We also had wooden carpet covered platform sections that we could build up to two tiers high to hold string sections. These were usually miked with three to five capacitor microphones connected in series or series/parallel! Remember we had only 12 inputs to work with at that time even with the brand new custom consoles built by Universal Audio. The brass section was often placed on risers also with the woodwinds placed at right angles to the brass at floor level and using one or two RCA 44bx ribbon mikes in the bi-directional mode. This prevented a lot of acoustic cross talk from the brass. There was also a fold up, free standing booth around the drum set.

Bill liked to group the piano, bass, drums and guitar in a tight little group, and he usually used an RCA 77DX planted in one of the sounding board holes in the piano. In those days we usually miked the guitar speaker with a Shure 555, 666 or 639 mike and sometimes "went direct" by connecting a mike transformer to the speaker terminals with alligator clips. The bass (in those days usually a standup bass) was miked with an Altec 639 in cardioid mode. Drums usually had one capacitor microphone over the top of the set, and sometimes, depending on the music, a Shure 555 on the kick drum and also under the snare.

This was pretty much the same for most sessions although the actual placement varied with the type of music and the size of the orchestra. The large studio could easily hold a full symphony orchestra, and sometimes did. The next smaller Studio B could easily hold 15 to 20 musicians. The smallest Studio C was large enough for a small Blues Band of about 7 to 10 people, but was used much more as an announce booth for Agency work. As previously stated, Bill liked to try out all the latest microphones, and his connections in the audio industry coupled with his reputation made it fairly easy to get "loaners" from the industry reps for "testing." In Hollywood we had a huge assortment of microphones to choose from and many times we had them all in use at one time and had to rent a few more! Fortunately most of the big studios in Hollywood at that time were very friendly with each other and many times borrowed equipment from each other!

**Bob:** Here we're discussing United and Western in Hollywood. Bones Howe, (who came to United from Radio Recorders, then became an independent producer and quite successful) wanted to do a session at United, using only RCA 77DX mikes. It took him some time to get them all, as they were a popular mike. As I recall, he was very pleased with the result.

United had two studios, A, and B. We've put pictures of the studios in the center section. They were not glamorous rooms, as the imaginative use of wood, fabrics and stone was not yet in use. The industry hadn't come that far. Remember, we're discussing 1959, 50 years ago!

Studio A could handle 60 musicians, plus a small choir. The room was roughly rectangular, with the control room at the northeast corner. A Steinway B grand piano was kept in regular tune by our regular piano tuner, Russ Reiner. (I think that was his name.) We didn't have mike lockers, in those days they weren't necessary. I don't recall we lost many mikes. The floor was asphalt tile over "thinset" cement, over one inch plywood flooring. The ballasts for the fluorescent lights were above the studio. There were double doors at the south end of the room, for bringing in percussion instruments, tympani and guest pianos. The walls were a sawtooth design, breaking up any standing waves.

Studio B was a good room, a pleasant balance between size and intimacy. The basic elements were the same as Studio A. The room was more rectangular, with the control room at the north end. There was a space in the flooring, filled with machinery isolating rubber which ran all around each studio between the double walls which isolated any impact sound (as from bass, or percussion) from being transmitted through the floor to adjacent areas, Especially since we were sharing the first floor with the film recording company next door.

Both rooms had the same basic design philosophy, but each room had its own character and sound. As you see from the pictures, the wainscoting was perforated Masonite over studs, with fiberglass insulation behind. This technique provided a fairly tough surface when instruments or mike stands hit the wall.

Both rooms sounded good, (but not necessarily the same) that was Bill's goal!

For our comments about Western Recorders and the three studios, go back to "1962," where Jerry and I discuss the building and the remodeling in detail.

## SOME THOUGHTS ABOUT THE OPERATION

**Bob:** From the beginning of United Recording in 1959 through 1970, we had a definite spirit of friendship throughout the studios. Some engineers worked very well with our clients, some preferred working behind the scenes. Some were excellent at relating to the technical nature of recording, others simply knew how to mix, not always understanding the acoustic nature of studios, nor the myriad collection of equipment available. Some read music and some didn't know how to read music.

But between Bill and the geniuses in the Traffic Department, the work got done, and very well. I think partly because United and Western were the "new guys" in town, and had to set an example. And partly because Bill wasn't simply the boss that you saw every so often, he was there in the midst of things.

If you wanted to try something new, Bill was more than agreeable. Paperwork? At a minimum. Documentation? Sometimes less than perfect. But somebody knew about the neat new devices, you just had to ask around. We had some great years, being part of this relatively new industry, catering to clients and talent that accepted us as a professional operation.

**Jerry:** We knew what we were doing most of the time and when we didn't, Bill was there to offer his ideas and suggestions. We were not afraid to innovate, as that was our stock in trade and it paid off handsomely. From studio procedures to lacquer mastering to film recording, we put new ideas to work for us and the clients loved it. We racked up an impressive list of "firsts" in the business and the word got out as the clients came in. The studios did well until some of our clients decided that they could do it themselves having made enough money to buy the equipment. Many found out that it wasn't enough. Many hired their own engineers, built their own studios and some succeeded. But the big studios took a hit from which many never recovered. The empire was crumbling!

## TIMELINE OF THE INDUSTRY

**1922** First recorded use of sound reinforcement in the US (and basic mixers) for inauguration of Warren G. Harding as US President.

**1930s** Beginning use of (probably) four channel mixers for dance band remotes, remote broadcasts. They were built by RCA, and referred to as OP-6 and OP-7 mixers. Recording studios, 6 input systems, not custom.

**1947** Carl Langevin was building individual components. RCA broadcast consoles were being modified for recording use.

**1948** Universal Recording, Chicago, built monaural consoles for their own use. Recording centers were Los Angeles, Chicago, Nashville, and New York.

**1955** Various binaural recordings made by individuals, among them Emory Cook, well known maker of record cutting heads. These records had two separate sets of grooves. One for left channel information, and another for the right. They were played using a two-headed tone arm.

**1958** Announcement by Westrex of *45/45* Stereo Disk Recording System.

**1958** United Recording in Los Angeles building custom consoles for their own use and sale (12 x 3). Probably the first use of equalizers in each microphone input position, pioneered by Bill Putnam.

**1958** Radio Recorders in Hollywood built Studio 10H with custom console (16 x 2). The console was built in house. They used Langevin EQ-251 equalizers in the microphone input positions.

**1958** Atlantic Records in New York purchased the third eight-track recorder using 1 inch tape from Ampex. It was called the 300-8. The first two machines were purchased by Les Paul. The machines had fixed azimuth heads.

**1959** First use of 3 channel 1/2" tape machines, boards modified from 2 buses to 3 buses.

**1964** Appearance of 4 channel consoles. What about Neve in the UK?

**1967** Electrodyne building stock and custom consoles with 4 busses. First use of summing amplifiers, developed by Don McLaughlin and John Hall.

**1970** Quad-Eight building stock and custom consoles, 4 buses.

**1971** 8 bus consoles appearing in studios.

**1972** Quadraphonic records were a hot item that year. JVC announced a discrete system, which allowed for four channels on the LP. That was one format. Sansui, Columbia, Electro-Voice (there may have been one more designer) all developed matrix systems. Unfortunately, they were not compatible with each other, much less with the discrete system. Since the separate designers could not agree on a single format, the equipment manufacturers eventually dropped the whole idea.

Tascam announcing 16 x 4 consoles for semi-pro use. By now, tape recorders are 16 track monsters utilizing 2" tape, and rapidly headed toward 24 track. There was even some 46 track (!) recording being done with the use of synchronizers to keep two 24 track machines in sync. It was now possible to assign a track to each member of an orchestra or group. This practically made recording engineers obsolete for original recordings, as he was only needed for the final mix-down or combination of all the tracks.

Groups could actually be recorded one person at a time, at different locations and even different days at their convenience. Likewise for the musicians in an orchestra. But it never really worked that way, because the performances lacked the spark only present when humans work and contribute together to a finished product. The results were inevitably flat and lifeless. There is a feeling of excitement, empathy and pride of accomplishment when musicians and artists gather together to produce an artistic product which is captured along with the performance. That feeling is somehow missing when the performers add their parts separately.

**1974** The creation of small studios (originally called 'project studios' became prevalent. Since the majority of recording was track by track (or 'layer cake' recording, the need for larger studios continually diminished. Mix Magazine regularly showcases new operations; one condition they all have in common is a smaller studio. With most of these, the control room is larger than the studios.

Scoring stages are still in demand in Los Angeles, New York, England, and Europe for feature motion picture sound tracks.

**The Present:** Digital Recording. Sounds are converted electronically into "bits" of 1's and 0's and recorded onto tape or more recently hard disks as magnetic pulses or pits blasted out of a substrate by lasers. Many critics say they can hear the discontinuities caused by the breaking of the natural analog world into bits and pieces, and abhor it. Many of these are still buying and playing their music on vinyl records, and indeed, a growing number of record companies are reverting back to vinyl albums albeit in very small numbers, many as special editions. Many musicians are staying with their analog equipment. Buying amplifiers that use tubes instead of solid state devices. Yes, they are still available! Playing vinyl records instead of Compact Discs, there are many music lovers who like the softer sound of analog. Recently a number of large chain stores in the country have begun to stock limited editions of vinyl LP records in some of their stores and they can be found in a number of specialty record shops in the larger cities. Another major indication that there may be a resurgence in the vinyl LP market is that Amazon.com has been quietly stocking up for the last few years. Now it only remains for the Record Companies to start releasing more of their albums on LP's again. The old saying "What goes around, comes around" certainly applies here, so don't throw out that old turntable and amplifier, you may want to use them again!

But the digital world is rushing forward like a freight train in the night engulfing everything analog from time to light and we might as well try to get used to it.

Like all of the electronic innovations before it, digital also has a finite lifetime. What will replace it has yet to be defined, but it **will** come and when it does, it will usher in a whole new generation of ways to capture and reproduce the sounds around us and create another new industry the likes of which we can only imagine!

## TROUBLE SHOOTING TIPS (EVEN FOR THE NON-TECHNICAL):

1. Using common sense is probably the best tip of all.
2. When examining a problem, or a possible problem, don't assume anything. And we do mean anything!
3. Even if you're not an engineer, you can still trouble shoot.
4. You're in the control room by yourself, and you have no sound, whereas you did five minutes ago, what do you do?
    - **a**. Does the system have power?
    - **b**. Check the settings on the input channel you're using. Are they OK?
    - **c**. If they are OK, look for the obvious items; is all equipment on?
    - **d**. Is the tape machine or computer running?
    - **e**. Are the VU meters reading? If so, did you hit an overall mute button?
    - **f**. If you didn't hit a mute button, are the monitor power amplifiers on?
    - **g**. If the VU meters aren't reading, punch up the test oscillator to a line input.
    - **h**. Do you read or hear the oscillator?
5. If you can, then check microphones and cables or the sound source.
6. We could go on, but this should give you a general idea.

Apart from our actual experiences working with the United family of companies, the following references are offered for your perusal:

United Recording Corporation published 21 newsletters through the years. You'll find those available as PDF files at:
http://studioelectronics.biz/index.html

Bill Putnam, Jr. (we use 'junior' only for purposes of clarification) has a neat website for Universal Audio at:
http://www.uaudio.com

There have been some books written about the recording industry. A search of the Internet or Amazon will provide some references. There are trade magazines available, "MIX," "Electronic Musician." "MIX" had a two-part article about Bill, with few errors.

CPSIA information can be obtained at www.ICGtesting.com
Printed in the USA
BVOW04s1223201214

380011BV00009B/290/P